CONTAMINANT HYDROLOGY

COLD REGIONS MODELING

CONTAMINANT HYDROLOGY

COLD REGIONS MODELING

Edited by

S. A. Grant
I. K. Iskandar

CRC Press
Taylor & Francis Group
Boca Raton London New York

CRC Press is an imprint of the
Taylor & Francis Group, an **informa** business

CRC Press
Taylor & Francis Group
6000 Broken Sound Parkway NW, Suite 300
Boca Raton, FL 33487-2742

© 2000 by Taylor & Francis Group, LLC
CRC Press is an imprint of Taylor & Francis Group, an Informa business

No claim to original U.S. Government works

ISBN-13: 978-1-56670-476-2 (hbk)

Library of Congress Card Number 99-086919

Visit the Taylor & Francis Web site at
http://www.taylorandfrancis.com

and the CRC Press Web site at
http://www.crcpress.com

Library of Congress Cataloging-in-Publication Data

Contaminant hydrology : cold regions modeling / S.A. Grant, I.K. Iskandar [editors].
    p. cm.
    Proceedings of a workshop held in Anchorage, Alaska in August 1995.
    Includes bibliographical references and index.
    ISBN 1-56670-476-6
    1. Groundwater—Pollution—Cold regions—Mathematical models—Congresses. 2. Groundwater flow—Cold regions—Mathematical models—Congresses. 3. Groundwater—Purification—Cold weather conditions—Congresses. 4. Hydrology—Cold regions—Congresses. 5. Cold regions—Congresses. I. Grant, S. A., 1953- II. Iskandar, I. K. (Iskandar Karam), 1938-
TD426 .C663 2000
551.49'0911—dc21
                                        99-086919
                                           CIP

Preface

Approximately 50% of the Earth's land mass is frozen at some time during the annual cycle: 20% of the land contains permafrost, and the other 30% is underlain with discontinuous permafrost or is subjected to several freeze/thaw cycles per year.

Terrestrial environmental contamination in cold regions is an increasing concern in many areas of the world because it affects some of the most traditionally pristine areas and because environmental cleanup in cold regions presents substantial operational difficulties that may increase costs considerably. Moreover, the extreme temperature range, soils and geology, the unique biological diversity, the freezing and thawing of pollutants, and the impact on human activities make environmental site assessments and remediation challenging tasks.

While much has been learned about contaminant fate and transport in cold regions, much more remains to be done, especially in understanding the effects of cold region environments and predicting the effectiveness of candidate remedial actions. Additionally, it is difficult for decision makers to keep abreast of the results of the most recent research, and thus they are less able to make cost-efficient and effective choices. This lag in understanding prolongs remediation of contaminated sites, extends the diversion of resources, and increases the final costs of cleanup.

Approximately 60 scientists and engineers from the United States, Canada, England, and Russia attended a workshop held in Anchorage, Alaska in August 1995. The objectives of the workshop were to

- provide a forum and direct communications between the users and developers of contaminant transport models;
- facilitate exchange of expertise between European and North American environmental scientists and engineers;
- define the status of cold regions contaminant transport models; and
- identify knowledge gaps and recommendations for basic research needs for cold regions contaminant transport.

The 14 chapters of this book constitute the proceedings of the workshop. Section I consists of four chapters that discuss the nature of contaminant hydrology in cold regions:

Chapter 1 provides an overview on problems of contaminant hydrology in Siberia.

Chapter 2 describes a direct measurement method for air distribution in soils.

Chapter 3 details the use of subsurface frozen barrier and recovery trench for contaminant removal.

Chapter 4 highlights strategies for development of cost-effective amelioration procedures for oil spills in cold regions.

Section II consists of five chapters that present example applications of models for cold regions:

Chapter 5 details basic guidelines for conducting groundwater modeling to meet environmental requirements.

Chapter 6 addresses the hydrogeological problems in developing the diamond-bearing deposits in northern regions of Russia.

Chapters 7 and 8 provide examples on the use of freezing to concentrate solutes in liquid radioactive waste and to contain radioactive wastes in permafrost, respectively.

Chapter 9 describes the use of a hierarchical neural network for interpretation of ground-penetrating radar and for permafrost and stratigraphic layer identification.

Section III comprises five chapters on development of models for cold regions:

Chapter 10 discusses the permeability of frozen silt to organic contaminants.

Chapter 11 presents a pore-scale model for soil freezing.

Chapter 12 presents simulation models that were recently developed to describe the fate or movement of chemicals in seasonally frozen soils. In addition, coupled water, heat, and solute transport models in unsaturated soils have been advanced. The effect of freezing and thawing processes on contaminated redistribution has also been addressed.

Chapter 13 provides a general modeling approach for heavy metals retention kinetics in soils.

Chapter 14 summarizes a field and modeling study of groundwater contaminant transport in a discontinuous permafrost region.

The workshop was cosponsored by the U.S. Army Cold Regions Research and Engineering Laboratory, the USAE Waterways Experiment Station, the U.S. Army Research Office, and the U.S. Army Material Command. Representatives from the U.S. EPA, DOE, USGS, USDA, AEC, U.S. Army Corps of Engineers, and several universities attended the workshop.

We wish to thank the authors for their contributions and time. Financial and logistic support were provided by the cosponsors. In particular, we wish to acknowledge the following individuals: Dr. Jerry Comati and Carole O'Connor of the European Research Office for facilitating the attendance of several professionals from outside the U.S.; Douglas Johnson and Crystal Fosbrook of the Public Works, U.S. Army, Alaska Command, and Mark Wallace of Alaska District, Corps of Engineers, for hosting the workshop in Anchorage; John Rouillad and his wife, Dianna, for providing logistics and field support for us and the participants. We gratefully acknowledge Donna Harp, Susan Hardy, David Cate, and Edmund Wright at CRREL for typesetting and technical editing.

S.A. Grant
I.K. Iskandar

The Editors

S.A. Grant received his Ph.D. in soil science from the University of Florida in 1987. He is a member of the Committee on Physicochemical Phenomena in Soils, Transportation Research Board, National Research Council; American Society of Agronomy; Soil Science Society of America; International Union of Soil Scientists, American Geophysical Union; and American Statistical Association.

Dr. Grant is the author or co-author of 19 papers. He has been a peer reviewer for *Advances in Water Resources*; *Cold Regions Science and Technology*; *ES&T*; *Industrial and Engineering Chemistry*; *Journal of Physical Chemistry*; *Journal of Geophysical Research — Solid Earth*; *Proceedings of American Society of Civil Engineers*; *Soil Science Society of America Journal*; *Transportation Research Record*; and *Water Resources Research*.

I.K. Iskandar received his Ph.D. degree in soil science and water chemistry from the University of Wisconsin, Madison, in 1972. He is currently a Research Physical Scientist at the Cold Regions Research and Engineering Laboratory (CRREL) and a Distinguished Research Professor at the University of Massachusetts, Lowell. During his tenure at CRREL, Dr. Iskandar developed two major research programs. The first, on land treatment of municipal wastewater, which he successfully coordinated and supervised for 9 years, concerned the research on transformation and transport of nitrogen, phosphorus, and heavy metals. The second program examined environmental quality in cold regions. In the early 1980s, Dr. Iskandar's research efforts were focused on the fate and transformation of toxic chemicals in soils, development of nondestructive methods for site assessments, and development and evaluation of *in situ* remediation alternatives. He was the first to propose the use of a frozen ground barrier for containment of toxic waste. He is a fellow of both the Soil Science Society of America (SSSA) and the American Society of Agronomy (ASA), a vice president of the International Society of Trace Elements Biogeochemistry, and a member of the International Union of Soil Science.

Dr. Iskandar has edited or co-edited 10 books. He has written more than 20 chapters of books; published more than 100 technical and reference papers and reports; presented more than 55 invited lectures, seminars and symposia; and made 45 other presentations.

Dr. Iskandar has organized and co-organized many national and international workshops and symposia. His numerous awards include the Army Science Conference, 1979; the Army Research and Development Award, 1988; CRREL Research and Development Award, 1988; and several exceptional performance awards from the U.S. Army Cold Regions Research and Engineering Laboratory.

Contributors

Lawrence J. Acomb
Geosphere, Inc.
3055 Seawind Drive
Anchorage, AK 99516, U.S.A.
+1 907-345-7596 (telephone)
+1 907-345-8066 (fax)

F.G. Atroshchenko
Institute of Environmental Geology
Russian Academy of Sciences
University Research Center of
  Environmental Hydrogeology
St. Petersburg State University
14-Line, Dom 29
199178 St. Petersburg, Russia
+7 812-321-9749 (telephone)
+7 812-327-4922 (fax)
fedor@hydra.nw.ru

John M. Baker
U.S. Department of Agriculture
Agricultural Research Service
Department of Soil, Water, & Climate
1991 Upper Buford Circle
St. Paul, MN 55108, U.S.A.
+1 612-625-4249 (telephone)
+1 612-625-2208 (fax)
jbaker@soils.umn.edu

James Baldwin
Foster Wheeler
  Environmental Corporation
601 Telegraph Canyon Road
Unit 116
Chula Vista, CA 91910, U.S.A.
+1 619-702-3942 (telephone)
+1 619-702-4113 (fax)
jbaldwin@fwenc.com

Ron Borrego
Harding Lawson Associates
707 17th Street
Suite 2400
Denver, CO 80202, U.S.A.
+1 303-293-6117 (telephone)
+1 303-292-5411 (fax)
rborrego@harding.com

Glenn R. Bruck
U.S. Environmental Protection Agency,
  Region 10
1200 Sixth Avenue
Seattle, WA 98101, U.S.A.
+1 206-553-0691 (telephone)
+1 206-553-0119 (fax)
bruck.glenn@epamail.epa.gov

I.V. Chernysheva
Vernadsky Institute of Geochemistry
  & Analytical Chemistry
19 Kosygin Street
Moscow 117975, Russia
+7 095-137-57-42 (telephone)
+7 095-938-2054 (fax)
elkor@geokhi.msk.su

Ann M. Farris
Environmental Engineer
Nortech Environmental and
  Engineering Company
P.O. Box 81727
Fairbanks, AK 99708, U.S.A.
+1 907-452-1884 (telephone)
afarris@mosquitonet.com

Lin A. Ferrand
Center for Water Resources
  and Environmental Research
Department of Civil Engineering
City University of New York (T-110)
New York, NY 10031, U.S.A.
+1 212-650-8017 (telephone)
+1 212-650-6965 (fax)
lf@ce-mail.engr.ccny.cuny.edu

Gerald N. Flerchinger
U.S. Department of Agriculture
Agricultural Research Service
800 Park Boulevard
Plaza IV
Suite 105
Boise, ID 83712-7716, U.S.A.
+1 208-334-1363 (telephone)
+1 208-334-1502 (fax)
gflerchi@nwrc.ars.pn.usbr.gov\

Steven A. Grant
U.S. Army Cold Region Research
  and Engineering Laboratory
72 Lyme Road
Hanover, NH 03755, U.S.A.
+1 603-646-4446 (telephone)
+1 603-646-4561 (fax)
steven.a.grant@usace.army.mil

Larry D. Hinzman
Professor of Water Resources
P.O. Box 755860
441E Duckering Building
Water and Environmental
  Research Center
University of Alaska Fairbanks
Fairbanks, AK 99775-5860, U.S.A.
+1 907-474-7331 (telephone)
+1 907-474-7979 (fax)
ffldh@uaf.edu

Scott L. Horwitz
Engineering Field Activity Northwest
Naval Facilities Engineering Command
19917 7th Avenue NE
Poulsbo, WA 98370-7570, U.S.A.
+1 360-396-0047 (telephone)
+1 360-396-0857 (fax)
horwitzsl@efanw.navfac.navy.mil

Iskandar K. Iskandar
U.S. Army Cold Region Research
  and Engineering Laboratory
72 Lyme Road
Hanover, NH 03755, U.S.A.
+1 603-646-4198 (telephone)
+1 603-646-4730 (fax)
iskandar@crrel.usace.army.mil

Ronald A. Johnson
Professor of Mechanical Engineering
P.O. Box 755905
337 Duckering Building
Mechanical Engineering Department
University of Alaska Fairbanks
Fairbanks, AK 99775-5905, U.S.A.
+1 907-474-6096 (telephone)
+1 907-474-6141 (fax)
ffraj@uaf.edu

Douglas L. Kane
Professor of Water Resources
  and Civil Engineering
P.O. Box 755860
539 Duckering Building
Water and Environmental
  Research Center
University of Alaska Fairbanks
Fairbanks, AK 99775-5860, U.S.A.
+1 907-474-7808 (telephone)
+1 907-474-7979 (fax)
ffdlk@uaf.edu

A.N. Kazakov
All Russian Research and Design
  Institute of Production Engineering
Ministry of Atomic Energy
33 Kashirskoye Shosse
Moscow 115 409, Russia
+7 095-324-6434 (telephone)
mankin@dol.ru

Igor L. Khodakovsky
Vernadsky Institute of Geochemistry
  & Analytical Chemistry
19 Kosygin Street
Moscow 117975, Russia
+7 095-137-57-42 (telephone)
+7 095-938-2054 (fax)
elkor@geokhi.msk.su

Greg J. Light
Alaska Department
  of Environmental Conservation
610 University Avenue
Fairbanks, AK 99709, U.S.A.
glight@envircon.state.ak.us

N.F. Lobanov
All Russian Research and Design
  Institute of Production Engineering
Ministry of Atomic Energy
Russian Federation
33 Kashirskoye Shosse,
Moscow 115 409, Russia
+7 095-324-6434 (telephone)
mankin@dol.ru

Reinhold Ludwig
Electrical and Computer
  Engineering Department
Worcester Polytechnic Institute
100 Institute Road
Worcester, MA 01609, U.S.A.
+1 508-831-5199 (telephone)
+1 508-831-5680 (fax)

Daniel J. McKay
U.S. Army Cold Regions Research
  and Engineering Laboratory
72 Lyme Road
Hanover, NH 03755, U.S.A.
+1 603-646-4738 (telephone)
+1 603-646-4561 (fax)
dmckay@crrel.usace.army.mil

Mikhail V. Mironenko
Vernadsky Institute of Geochemistry
  & Analytical Chemistry
19 Kosygin Street
Moscow 117975, Russia
+7 095-137-2484 (telephone)
+7 095-938-2054 (fax)
mironenko@geokhi.ru

Valery A. Mironenko
Institute of Environmental Geology
Russian Academy of Sciences
University Research Center of
  Environmental Hydrogeology
St. Petersburg State University
14-Line, Dom 29
199178 St. Petersburg, Russia
[deceased January 26, 2000]

Oleg S. Pokrovsky
Laboratoire de Géochimie,
CNRS URM 5563
Université Paul Sabatier
38 rue des Trente Six Ponts
31400 Toulouse, France
+33 05 61 55 84 05 (telephone)
+33 05 61 52 05 44 (fax)
oleg@lucid.ups-tlse.fr

W. Gareth Rees
Cambridge University
Scott Polar Research Institute
Lensfield Road
Cambridge CB2 1ER, U.K.
+44 1223 336575 (telephone)
+44 1223 336549 (fax)
wgr2@cus.cam.ac.uk

Dmitry V. Repin
Electrical and Computer
  Engineering Department
Worcester Polytechnic Institute
100 Institute Road
Worcester, MA 01609, U.S.A.
+1 508-831-5199 (telephone)
+1 508-831-5680 (fax)

Dan W. Riseborough
Geotechnical Science Laboratories
Carleton University
1125 Colonel By Drive
Ottawa, Ontario
K1S 5B6, Canada

Amadeo J. Rossi
CH2M Hill Constructors Incorporated
P.O. Box 91500
Bellevue, WA 98009-2050, U.S.A.
+1 425-453-5005 ex 5299 (telephone)
+1 425-462-5957 (fax)
arossi@CH2M.com

H. Magdi Selim
Department of Agronomy
Louisiana State University
Sturgis Hall
Baton Rouge, LA 70803-2110, U.S.A.
+1 225-388-1332 (telephone)
+1 225-388-1403 (fax)
mselim@agctr.lsu.edu

A.I. Shapkin
Vernadsky Institute of Geochemistry
  & Analytical Chemistry
19 Kosygin Street
Moscow 117975, Russia
+7 095-939-7082 (telephone)
+7 095-938-2054 (fax)
ashapkin@geokhi.ru

Egbert J.A. Spaans
Escuela de Agricultura de la Region
  Tropical Humeda (EARTH)
Apartado 4442-1000
San Jose, Costa Rica
+506 255-2000 ext. 3113 (telephone)
+506 255-2726 (fax)

John M. Sullivan, Jr.
Electrical and Computer
  Engineering Department
Worcester Polytechnic Institute
100 Institute Road
Worcester, MA 01609, U.S.A.
+1 508-831-5199 (telephone)
+1 508-831-5680 (fax)
sullivan@wpi.edu

Oleg F. Vasiliev
Institute of Water
  and Environmental Problems
Siberian Branch of the Russian
  Academy of Sciences
Morskoy Prospect 2
630090, Novosibirsk, Russia
+7 383-2-30-20-05 (telephone)
+7 383-2-30-20-05 (fax)
vasiliev@iwep.nsk.su

T. Les White
Geotechnical Science Laboratories
Carleton University
1125 Colonel By Drive
Ottawa, Ontario
K1S 5B6, Canada
+1 613-520-2600 ext. 8297 (telephone)
+1 613-520-3640 (fax)

Kim Winnicky
SEACOR Environmental
  Engineering Inc.
#9 6421 Applecross Road
Nanaimo, British Columbia
V9V 1N1, Canada
+1 250-390-5050 (telephone)
+1 250-390-5042 (fax)

Peter J. Williams
Geotechnical Science Laboratories
Carleton University
1125 Colonel By Drive
Ottawa, Ontario
K1S 5B6, Canada
+1 613-520-2852 (telephone)
+1 613-520-3640 (fax)
pwilliam@ccs.carleton.ca
pjw1005@cus.cam.ac.uk

Contents

SECTION I

The Nature of Contaminant Hydrology
and Case Studies for Assessment

CHAPTER 1

Problems of Contaminant Hydrology in Siberia

O.F. Vasiliev

CONTENTS

INTRODUCTION

    Siberia has the largest area and the richest natural resources in Russia. It covers a great part of North Asia and stretches for more than 7000 km from the east slope of the Ural Mountains to the mountain range of the Pacific Ocean watershed (the mountains confining the Kolyma Basin) in the east, and for about 3500 km from Cape Chelyuskin in the north to the Altai Mountains in the south. A comparatively small southern area of the West Siberian Lowland lies in the territory of the Kazakhstan Republic. This part of Kazakhstan (Northeast Kazakhstan, parts of Semipalatinsk, and Pavlodar regions) belongs to the single natural and geographical complex of Siberia. The Siberian region covers somewhat above 10 million km^2 or around 7% of the land territory of the Earth (Mikhailov, 1976).

Geographically, Siberia is situated in the middle northern latitudes and partially in the high northern latitudes, within the temperate and cold climate zones, a fact that governs its natural features. The southern mountain borders favor the formation of a sharp continental climate over most of the territory. The most distinctive feature is the long winter period, with very low winter temperatures. The average annual temperature almost everywhere in Siberia is below 0°C. As a consequence, about 70% of the region falls into the permafrost zone.

HYDROLOGY

Precipitation is distributed rather unevenly over the territory (from 100 to 2500 mm year^{-1}, decreasing toward the east) and over the seasons (Mikhailov, 1976). The common characteristic of the Siberian geographical structure is its hydrographic unity and the unique hydrologic regime of rivers. The overwhelming majority of rivers belong to the Arctic Ocean basin, and only a small part of the Transbaikalian region is drained by tributaries of the Amur River. The Ob, Yenisei, and Lena, which are among the world's largest rivers with a total annual runoff of about 1500 km^3, constitute the basis of the river network in Siberia. As for the total annual runoff of all Siberian rivers, it is somewhat above 2500 km^3, i.e., approximately 7% of the overall flow of the Earth's rivers. Most Siberian rivers are fed by thawed snow and ice in spring and by rains in summer and autumn. The groundwater feeding is usually small, often below 10%. For almost all the rivers, 80 to 90% of the annual flow takes place during the warm periods and no more than 7 to 15% in winter. The major flow occurs at the high water periods, which are in late spring in the southern and middle regions of Siberia and in early summer in the north. In Siberia, almost 90% of country settlements and large cities are located in the river valleys (Mikhailov, 1976).

TOPOGRAPHY

The significant extent of Siberia from north to south and from east to west causes a great diversity of natural landscapes. The overall surface area is 14% tundra and forest-tundra biomes, 42% taiga, 5% forest-steppe, 4% steppe, and 35% mountain and hills (Koropachinskii et al., 1994). From the latest data, the forest coverage is estimated to be 3300×10^3 km^2 (Sokolov et al., 1994).

A great part of Siberia (in particular West Siberia) is covered by wetlands with bogs. The area of wetlands in West Siberia is estimated as 800×10^3 to 900×10^3 km^2, including about 400×10^3 km^2 of peat bogs (Pomus, 1971; Neishtadt, 1971; Kats et al., 1963). The total area of wetlands in Siberia is about 2000×10^3 km^2.

DEMOGRAPHY

A low population density and a very nonuniform distribution of the population over the territory are the distinctive demographic features of the Siberian region.

There are only 24 million people throughout this extended area. Only the agricultural forest-steppe band of West Siberia and Kuzbass contains comparatively populous regions (40 to 60 people per km^2). A high concentration of industrial enterprises in a number of large industrial centers and the prevalent superlarge enterprises characterize industrial progress in Siberia (Mikhailov, 1976).

ENVIRONMENTAL CONCERNS

Because of the intensive development of various branches of economy in Siberia, in particular in its northern regions, many complicated environmental issues have arisen. These are related to water resources management and protection, as well as to usage and disturbance of natural systems, including those which are closely and inherently connected with hydrologic systems.

Ecological well-being of the cold regions is heavily affected by the chemical contamination of hydrologic systems, including watersheds, surface water, and groundwater.

The contamination of hydrologic systems in Siberia has different characteristics and scales in various cases. The response of natural systems (ecosystems), whose functioning is tightly connected with water cycling (or water exchange processes), depends very much upon the stability in different landscape zones (tundra and forest-tundra, taiga, peat bogs, etc.). Occurrence of permafrost plays a significant role. As is well known, cold-region ecosystems with their high vulnerability and low regenerating capability are especially sensitive to direct and indirect human impacts.

Industrial development in Siberia gave rise to various kinds of environmental damage and, sometimes, even devastation. The damage includes distortions in the behavior of hydrologic and hydrogeologic systems, their qualitative state, and in the stability of landscapes, particularly in permafrost areas. Vast areas of Siberian territory have been subjected to environmental changes related to the contamination caused by environmentally detrimental technologies and engineering equipment.

There are many examples of large-scale local contamination of lands and waters in Siberia, because of various industrial activities such as oil and natural gas extraction, processing of mineral, wood, and other natural resources, and use of transport systems. Oil refineries and metallurgical plants (ferrous and nonferrous) contribute to both direct and indirect contamination of watersheds and hydrologic systems due to discharge of wastewaters, emission of gaseous effluents (airborne pollutants), and disposal of wastes in tailing dumps.

THE OB-IRTYSH RIVER SYSTEM

Large-scale pollution of waters and soils occurs in the Ob-Irtysh River basin, one of the largest river basins of Siberia and the world. Some small and midsize tributaries of these large rivers, such as Inya, Tom, Chulym, Vakh, and Ishim, are highly contaminated, especially in the areas of industrial activities. While earlier pollution of river waters took place mainly in the upper stretches of the Ob-Irtysh River system (in the most populated and industrialized southern regions of the West Siberia), during the last 30 years the areas of industrial and urban pollution on the West Siberian territory expanded a great deal and reached the northern regions. That is the negative result of the development of the West Siberian oil and gas production complex. Therefore, the problem of oil contamination of watersheds (covered with enormous peat-bogs), rivers, and their floodplains in the Middle Ob River basin has become most urgent today.

We shall give some figures to better envisage the scales of the environmental impact of this industrial complex. There are almost 200,000 km of pipelines used in the regions of Tyumen and Tomsk. The pipelines have been constructed since the mid-1960s and they have been used under heavy technological and environmental conditions. To intensify oil production, water is often injected into an oil formation to increase reservoir pressure. A negative impact of this is the introduction of brine waters from aquifers into the system, which leads to the intensive corrosion and rapid wear of pipes. As a result, up to 2500 to 3000 pipe-break accidents take place per year, resulting in oil losses estimated at 1 to 2% of the total production (Mikhailova, 1995). As a consequence of these failures, crude oil and brine waters are released in very large quantities on watersheds and into water bodies. The region's flat relief, an abundance of wetlands, bogs and lakes, low gradients of streams, a cold climate, and perma-frost aggravate environmental consequences.

The accidents involving the oil pipeline network, wells, and pumping stations are the main source of contamination of watersheds and water bodies by oil hydrocarbons and brine salts. The use of slime storage pits at well sites, as well as the release of oily wastes from the technological systems, also plays an essential part in the contamination of hydrologic systems by oil hydrocarbons. The small rivers and some tributaries (e.g., the Vakh) of the Ob River, which are situated on (or contain) oil fields, are subjected to a considerable degree of the contamination.

Pollutants transferred by air contribute to the contamination of extensive terri-tories. The contribution of airborne pollutants (including hydrocarbons) in the con-tamination of watersheds is not estimated well. One of the sources of such a contamination is flaring of associated gas at gas separator outlets, in particular the emissions of the airborne dripping oil remaining after incomplete burning. Aromatic hydrocarbons play a significant part in oil contamination of water bodies and snow cover (Mikhailova, 1995).

All these result in the large-scale and substantial contamination of soils and waters by oil hydrocarbons in the Middle Ob River basin. The salinization of river waters by the brine salts amplifies the environmental impact.

The chemical contamination of the rivers in the Tyumen North and the rivers' low ability for self-purification have already brought about negative conse-quences for the aquatic ecosystems. The most visual of them is the damage of fisheries in the Middle and Low Ob River. The following data for the fishing harvest in the Tyumen region (according to the data by Mikhailova, 1995) give evidence to it:

Average harvest in years 1940 to 1964 (32×10^3 t/year)
Estimate of harvest at present (16 to 17×10^3 t/year)

Thus, the decline is about 40 to 50%.

THE TOM RIVER BASIN

Another example of large-scale environmental contamination is the situation in the area of Kuzbass Industrial Complex, which is located in the Tom River basin (a right tributary of the Ob River). In this case, a complicated environmental situation has resulted from the fast development of a number of industrial activities, including coal mining and processing of ferrous and nonferrous ores and other natural resources. Together with the growth of urban settlements, the industrial activity and lack of measures for environmental protection have caused substantial pollution of air, lands, and waters.

The Tom River is 827 km long and drains a watershed area of 62,000 km². The total mean annual runoff of the river is 34.1 km³/year (1080 m³/s). The basin contains several environmentally devastated urban and industrial areas of the Kuzbass Industrial Complex, which is the most important center of coal mining in Russia and one of the centers of metallurgical and chemical industries located in such cities as Kemerovo, Novokuznetsk, and others. The Kemerovo region is one of the most densely populated and heavily industrialized areas in Siberia, with well-developed supporting agricultural activities. In that region, some urban areas, industrial sites, and stretches of the Tom River and its tributaries are badly contaminated.

The Tom River system is the main source of water supply for the Kuzbass region and the ultimate collector for the industrial, municipal, and agricultural wastewaters. Over 1000 industrial plants and human settlements discharge incompletely treated wastewaters into the Tom River and its tributaries. The washout of the pollutants from the municipal territories and industrialized areas, wastes from the cattle-breeding complexes, and the pollutant runoff with surface waters from the agricultural lands are also very substantial. In addition, our recent data of field observations suggest that airborne pollutants play an essential role in contaminating the watershed area and its surface waters. As a result, the river waters contain practically all types of pollutants (a variety of organic compounds, heavy metals, and others).

Although the Tom River basin is situated in a southern part of West Siberia, the climate of this area is typically continental with a long and cold winter period. That gives the river a hydrologic regime with a very nonuniform seasonal distribution of runoff. The river reflects its mixed origins where input of snow melting prevails (during the period of spring flood, about 70% of total runoff is released). At the same time, winter is characterized by very low discharges (because of limited groundwater feeding); the part of the total runoff that is related to this period is only about 5%.

Such hydrological conditions are most unfavorable both for water supply (lack of water and low levels) and for the quality of river water. The latter is due to the fact that the concentrations of a number of point-source pollutants reach high levels because of low dilution.

A project has been proposed aimed at mitigating the water quality state and the hydrological regime along the lower part of the river: near Kemerovo and downstream. The concept of the project was to create a rather large river reservoir by constructing a dam near Krapivino, a settlement located upstream from Kemerovo, and to regulate river flow (runoff) providing sufficient discharges to dilute the pollutant load at the low-water periods. Construction work started in 1975, but in 1989 was stopped according to a governmental decision: the project had to be considered once more for a new environmental impact assessment (Vasiliev, 1998).

From 1990 through 1992, the Institute for Water and Environmental Problems and other institutes of the Siberian Branch of the Russian Academy of Sciences have carried out a research program, the main objectives of which were assessing the water management and ecological state of the Tom River basin and the engineering measures proposed for the improvement of water resources and quality management. This wide-ranging multidisciplinary study has included the collection and analysis of available hydrological, hydrochemical, ecological, and other data on the river system and its watershed area, and field investigations for enlarging this information and assessing environmental impacts of the Krapivino project. The field investigations have been pursued mainly by an extensive survey of hydrochemical, biogeochemical, and hydrobiological data specifying the present state of the Tom River basin: quality of surface and groundwaters; pattern of sediment, soil, and snow pollution; and species diversity under conditions of anthropogenic stress and bioaccumulation of chemicals by the river inhabitants.

The hydrochemical study of water quality reveals that the Tom River and some of its tributaries are heavily exposed to anthropogenic contamination, especially immediately downstream from the large industrial centers. Main pollutants are the organic compounds (petroleum compounds, phenols, polycyclic aromatic hydrocarbons, formaldehyde, aniline, organic chlorine compounds, some amines, naphthalene and its derivatives, and dibutylphthalate and its derivatives), nitrate and ammonia nitrogen, as well as some heavy metals (cadmium, zinc, chromium, copper, etc.). Concentrations of the above substances often exceed drastically the national standards for water quality in natural water bodies.

The pattern of water contamination of the river changes over the annual hydrological cycle. In winter low water, the pollution by petroleum compounds is dominant everywhere along the river. Some stretches of the river are highly polluted by volatile phenols and ammonia nitrogen. Although the spring floodwater dilutes the petroleum compounds, water pollution by pesticides increases everywhere along the river. Also, the level of contamination by formaldehyde, phenols, and nitrate increases locally during the high water period in spring.

As a rule, heavy metals content in the water and bottom sediments does not exceed the normal background. However, there are contaminated stretches near industrial centers (e.g., Novokuznetsk and Kemerovo).

SOURCES OF CONTAMINATION

Currently, the quality of surface waters in the Tom River basin is determined to a great extent by pollution from different sources, including industrial plants, urban sewerage systems, agricultural lands, and urban and industrial areas. Therefore, its essential improvement is impossible without significant reduction of quantities of contaminants emission to air and of water and soil pollution from industrial and urban sources, as well as more careful use of fertilizers, pesticides, and other agrochemicals.

Research was aimed at establishing the scientific basis to develop a strategy for improving the environmental state of the Tom River and its basin. The results obtained thus far have demonstrated conclusively that the water quality problem cannot be considered separately from other environmental issues of the region. It can be done only within the total environmental context of the river basin. There is a need for developing a strategy of integrated management of water and environmental quality on the river-basin scale (Vasiliev, 1998). As far as the fate of the Krapivino reservoir project is concerned, there is not yet a final decision.

An example of industrial impact in the northern environment is that from the Norilsk Mining and Smelting Enterprise, the activity of which is based upon the large-scale exploitation of sulfide copper–nickel ore deposits. The mining technology includes the use of open-cast mines in the permafrost area. Therefore there are two sources of atmospheric pollution here: emissions of gaseous effluents (in particular, sulfur dioxide SO_2) from smelters and weathering of dust from spoil heaps. The deposition of dust and aerosols on the watershed surface results in the substantial contamination of soils, vegetation, snow cover, and waters over a large territory. A similar situation takes place in some other areas of the mining industry in Siberia.

Deep open-cast mines are used for the extraction of diamonds in Yakutia (in the permafrost area as well) at Mirnyi and to the north of Mirnyi in the area of Aikhal. In addition to the air pollution and the disturbance of groundwater regimes, an environmental problem has arisen at Mirnyi because of groundwater drainage: pumping of brine waters from deep aquifers to the Vilyui River resulted in the salinization of river waters.

Gold mining with the use of mercury technology (though the use of mercury is limited now, it is still in use in Siberia) is another example of contamination of river hydrologic systems. Hydrologic processes, including sediment and suspended matter transport, play a governing role in the mercury migration in a fluvial system. There are many river basins in Siberia, in particular in its eastern regions, where the gold mining is an important sector of industrial activities.

Of special interest are the hydrochemical regimes and the behavior of contaminants in river reservoirs situated in the permafrost areas. There are three in Siberia on the rivers: Vilyui (Lena River basin), Khantaika (Yenisei River basin), and Kolyma (East Siberian Sea basin). The Kolyma Reservoir has recently been created in mountainous conditions (the depth near the dam is 120 m). The river basin is under environmental stress caused mainly by the intensive gold mining in this area. Therefore there is a problem of the reservoir being contaminated by the wastes of mining activity and by municipal sewage. The hydrochemical regime of the reservoir, including the issue of heavy metals contamination, is considered in the recent paper by Susekova and Oganesyan (1996).

The examples of environmental contamination problems given above for the Siberian conditions attest that development of contaminant-transport models for hydrologic systems that include watersheds, surface waters, and groundwaters in the northern areas is of significance today. The transfer and chemical transformation of hydrocarbons, especially aromatic ones, in the hydrologic systems of cold-climate regions are a modeling task of particular priority.

The following knowledge gaps and research needs merit special attention:

- Role of surface and groundwater interaction in contaminant transport processes, coupling contaminant transport models with those for surface/groundwater interaction
- Modeling of contaminant transport and water quality in river basins with wetlands (bogs) and permafrost areas
- Role of ice cover in contaminant-transport processes in rivers and water bodies, and oxygen exchange under ice cover

REFERENCES

Kats, N. Ya. and M.I. Neishtadt. Wetlands, *West Siberia,* Izdatelstvo AN SSSR, Moscow, 230–249, 1963 (in Russian).

Koropachinskii, I.Yu. and V.P. Sedelnikov. Plant Resources of Siberia: Their Current State and Protection, *Sibirskii Ecologicheskii Zh.* 1, 17–28, 1994 (in Russian).

Mikhailov, N.I. *Siberian Nature. Geographical Aspects.* Mysl, Moscow, 1976 (in Russian).

Mikhailova, L.V. Chemical Contamination as One of the Major Ecological Problems in the Ob-Irtysh Region, in *The Ways and Means of Attaining Balanced Ecological and Economical Development in West Siberia Oil Region: Proceedings of Conference, Nizhnevartovsk, 1994.* Nizhnevartovsk, IPP "Uralskii rabochii," 1, 43–46, 1995.

Mikhailova, L.V. Present-Day Ecological Situation in Middle Ob River Basin, in *Hydrological and Ecological Processes in Reservoir and Its Watershed Basins: Proceedings of International Conference, Novosibirsk, Russia, 1995.* Novosibirsk, 137–138, 1996.

Neishtadt, M.I. World Natural Phenomenon — Bogging Up of West Siberian Plain, *Izvestiya AN SSSR, Seriya Geographiya,* 1, 21–34, 1971 (in Russian).

Pomus, M.I. (Ed.). *Soviet Union. Russian Federation. West Siberia.* Mysl, Moscow, 1971 (in Russian).

Sokolov, V.A., A.S. Atkin, and I.V. Semechkin. Wood Resources of Siberia, *Sibirskii Ekologicheskii Zh.,* 1, 39–46, 1994 (in Russian).

Susekova, N.G. and A.Sh. Oganesyan. Hydrochemical Regime of Kolyma Reservoir on Different Stage of Its Filling, *Vodnye Resursy.* 23(3) 351–360, 1996 (in Russian).

Vasiliev, O.F. Water Quality and Environmental Degradation in the Tom River Basin (Western Siberia): the Need for Integrated Management Approach, in *Restoration of Degraded Rivers: Challenges, Issues and Experiences.* Loucks, D.P., Ed., Kluwer Academic, Dordrecht. 283–292, 1998.

CHAPTER 2

Direct Measurements of Air Distribution during *in Situ* Air Sparging

Daniel J. McKay and Lawrence J. Acomb

CONTENTS

INTRODUCTION

In situ air sparging (IAS) is popular terminology denoting the injection of air into saturated soils to induce remediation. Site-specific designs of these air-based systems are complicated by uncertainties with respect to the local geology, spatial variability, and lack of a well-established connection between theory and engineering practice. Consequently, pilot tests are typically performed to evaluate potential performance and establish the relationship between injection pressure, flow rate, and radius of influence (ROI) for a particular geological setting (Marley et al., 1992; Marley et al., 1994; Martin et al., 1992). Conventional measurements to establish the ROI are water table mounding, soil gas pressure, dissolved oxygen, and soil gas composition (Brown et al., 1994; Brown and Jasiulewicz, 1992; Johnson et al., 1993). However, these data appear to be inadequate for quantitative evaluation of the IAS process (Hinchee, 1994). A study of the traditional techniques indicated "erroneous impressions" of the extent of saturated zone air flow by a factor of at least 2 to 8 when compared with the actual distribution determined by cross-borehole electrical resistance tomography (ERT) (Lundegard and LaBrecque, 1995). The likelihood for misinterpretation of data can thus be minimized by utilizing direct measurements of the air distribution below the water table. In addition, a direct approach enables study of air flow characteristics not otherwise evident with indirect data.

This chapter describes the application of a neutron moisture probe to measure the changes in saturated-zone moisture content during operation of an IAS system. Decreases in moisture content are a result of groundwater displacement by the injected air. By monitoring the moisture content at a variety of locations around a sparge well, profiles of steady-state air distribution were developed. In addition, measurements were performed at discrete time intervals following system startup to evaluate transient responses.

SITE DESCRIPTION

The IAS system was constructed on a coastal island near Cordova, Alaska. The soil formation is composed of visually uniform, fine eolian and beach sands with a porosity of 42 to 44% as determined from dry bulk density and specific gravity measurements. Approximately 94% by weight consists of grain sizes less than 0.250 mm and 16% is less than 0.150 mm, while only 3% is less than 0.075 mm. The radial and vertical components of air-phase permeability were determined (Baehr and Hult, 1991) as 1.92×10^{-7} and 5.32×10^{-8} cm^2, yielding a 3.6:1 preference for horizontal vs. vertical air flow. The area is contaminated with 15,000 mg/kg (typical) of diesel-range organics (DROs), predominantly in the depth interval of seasonal water table fluctuation.

The site plan is shown in Figure 2.1. Sparge wells were constructed of 10.2-cm-diameter PVC with the top of a 1.52-m length of screen length located about 7 m below the ground surface. A row of neutron-probe access pipes was installed between sparge wells SW04 and SW07 and identified as NP01 to NP06. These were constructed of black iron pipes with a 5.1-cm diameter and a wall thickness of 3.91 mm. A stainless steel access pipe with a 5.1-cm diameter and a wall thickness of 1.65 mm was installed adjacent to sparge well SW01 and identified as NP12. All probe access pipes were installed with a hollow-stem auger and are surrounded by formation soils that were washed or collapsed into place.

APPROACH

A model 501DR neutron moisture/density borehole probe (Campbell Pacific Nuclear) was used to measure changes in water content within an estimated radius

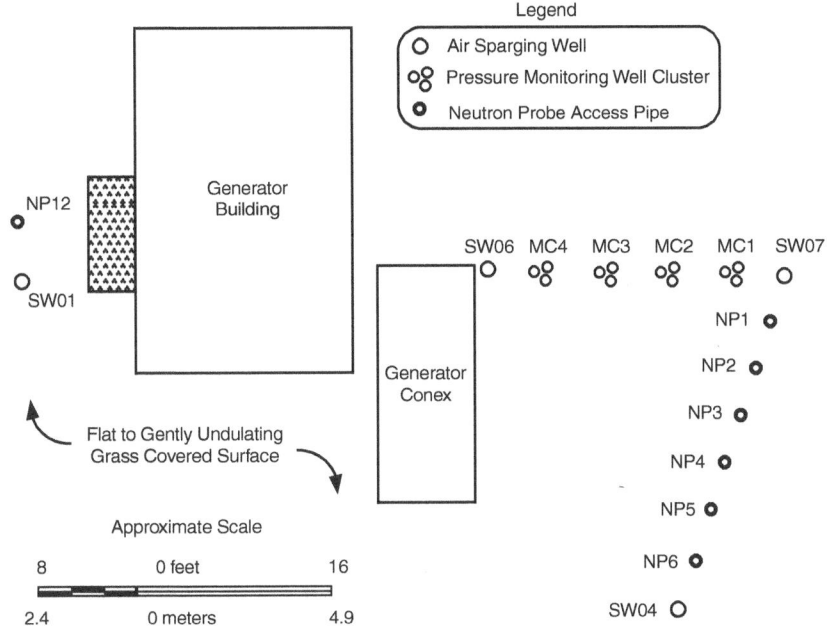

Figure 2.1 Location of air sparging wells and neutron-probe access pipes.

of 15 to 20 cm around the access pipes. Measurements were taken at 0.61-m intervals starting at the water table to about 8.5 m below ground surface, utilizing 64-s or 32-s counts such that a borehole could be profiled in about 6 to 12 min.

Baseline counts were performed at each measurement location prior to starting the air sparging system. These were assumed to represent 100% fluid-saturated values. The percentage of air saturation during air sparging, A_{ijt}, was then calculated as

$$A_{ijt} = \frac{(B_{ij} - C_{ij})}{B_{ij}} \times 100 \tag{2.1}$$

where C_{ijt} is the measurement at time t, location i, and depth j, and B_{ij} is the corresponding baseline measurement. Note that Equation 2.1 expresses the air content on a volumetric basis. The content of air on a mass basis varies proportionately with the hydrostatic pressure.

Counting data typically have a Poisson distribution. The estimate of uncertainty based on 20 measurements and a 0.025 significance level was 0.66%. By the method of Kline and McClintock (1953), the uncertainty in a comparison of two sets of 10 counts at a 0.025 significance level was determined to be 1.5% when Equation 2.1 is used. Not included in this analysis, however, are the effects of residual petroleum. Water and petroleum both contain hydrogen nuclei, which thermalize the high-energy neutrons during a count. Since petroleum hydrocarbons are less easily displaced by the injected air, their presence may cause one to underestimate the potential for air saturation at a given point.

RESULTS

Transience

Sparge well SW04 was operated at 19 m³/h, while measuring the air saturation in access pipe NP6 (1.1-m distance) at elapsed times ranging from 15 to 135 minutes following startup. Development of the air distribution following startup is shown in Figure 2.2. The highest measured value of air saturation, 34% at 1.3 m below the water table (BWT), occurred between 27 and 39 minutes, whereas the largest distribution of air is observed between 15 and 27 minutes following the onset of air flow.

Sparge well SW07 was operated at 13.6 m³/h and monitored after 0.25–0.5, 1.0–1.5, and 11 hours of the start of air injection. Measurements were made in NP1, NP2, NP3, and NP4, located 0.9, 2.2, 3.5, and 4.9 m from the sparge well, respectively. Figure 2.3A shows the results of the first set of readings along with

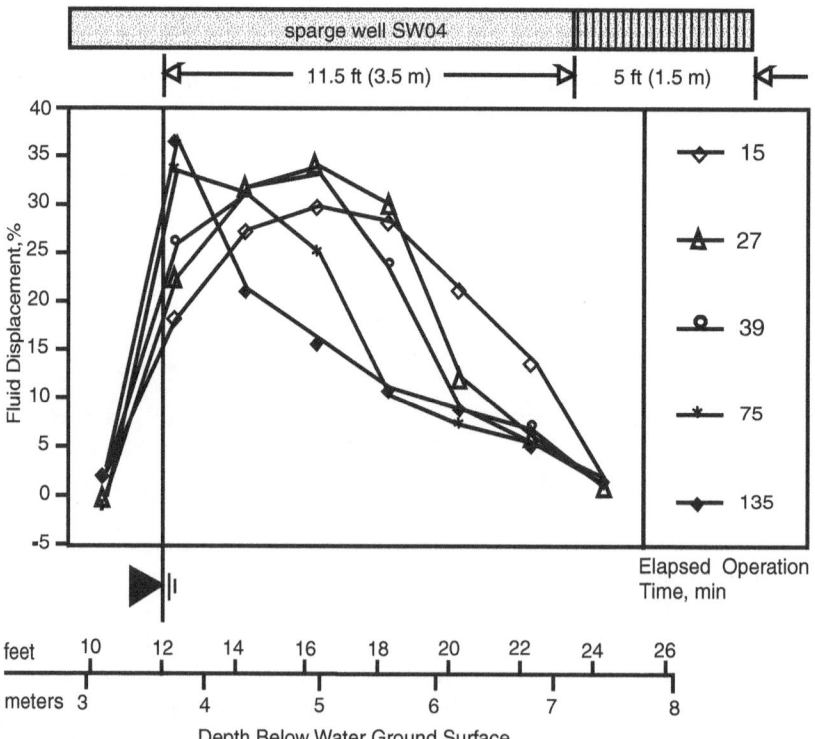

Figure 2.2 Air saturation in saturated zone during air injection in sparge well SW04. Values were determined by neutron moisture probe monitoring in pipe NP06.

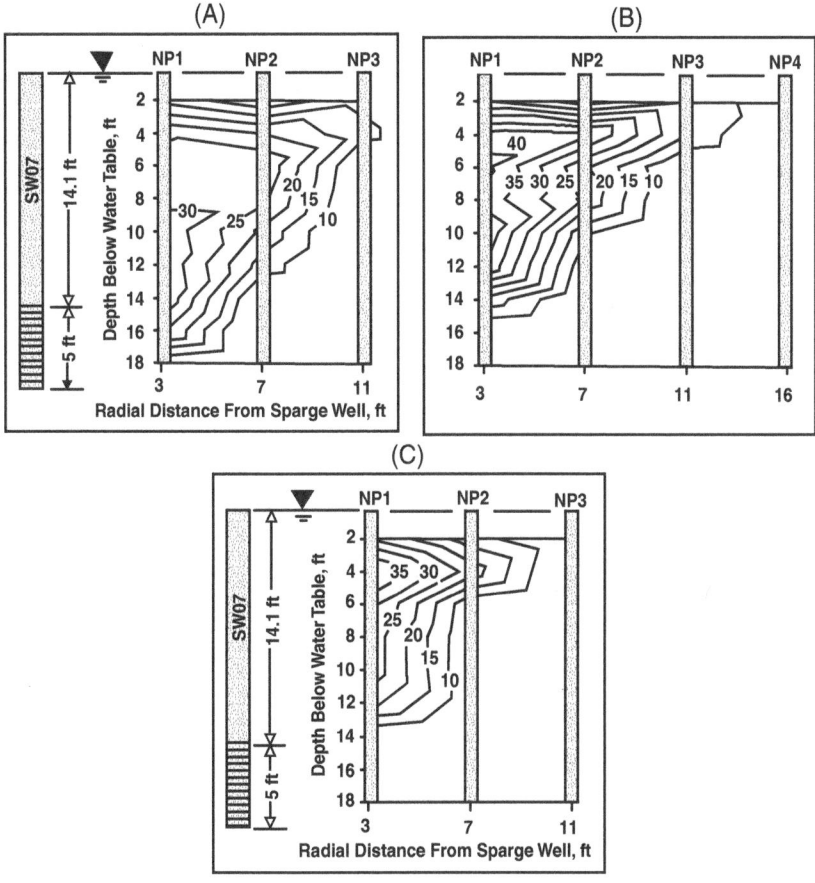

Figure 2.3 Air distribution profiles (percentage of pore volume containing air) during development stage of operation. Airflow rate was 13.6 m³/mine at sparge well SW07. (A) Expansion of phase (0.25–0.5 hours); (B) Near peak of expansion phase (1.0–1.5 hours); (C) Steady state (11 hours).

software-generated (Deltagraph® 3.5) contours in the plane of the measurements. The extent of the 10% air saturation contour is 3.6 m from the sparge well at about 1.2 m BWT and 5.3 m BWT at a radial distance of 0.9 m. The conical shape of the profile is characteristic of the expansion phase and was observed by Lundegard and Anderson (1993) in computer simulations and field measurements using electrical resistance tomography. The second set of readings are shown in Figure 2.3B. The 10% contour now extends laterally to 4 m from the sparge well at 1.2 m BWT, while shifting about 0.7 m upward at 0.9 m from the well. Since consolidation is already beginning at the deeper elevations, the air distribution is approaching the peak of the expansion phase. Measurements following 11 hours of operation, shown in Figure 2.3C, represent the steady-state profile. The 10% contour at 1.2 m BWT only extends 3 m from the sparge well and has shifted upward to

4 m BWT at 0.9 m from the well. Except near the screen and the water table, vertically oriented contours were characteristic of steady-state operation, as they always appeared following prolonged air injection.

Steady State

Steady-state profiles of air distribution at injection rates of 6.8, 13.6, 20.4, and 27.2 m³/h are shown in Figures 2.4A–D. The radial extent of the 10% contour (vertically oriented portion) is shown to increase by only 0.2 m when the flow rate was doubled from 13.6 to 27.2 m³/h. However, the level of air saturation inside of this region increased notably in proportion to flow rate. The elevation at which the 10% contour began to extend laterally near the water table was essentially constant (1.8 m BWT) between 13.6 and 27.2 m³/h. The depth of 10% air saturation at a

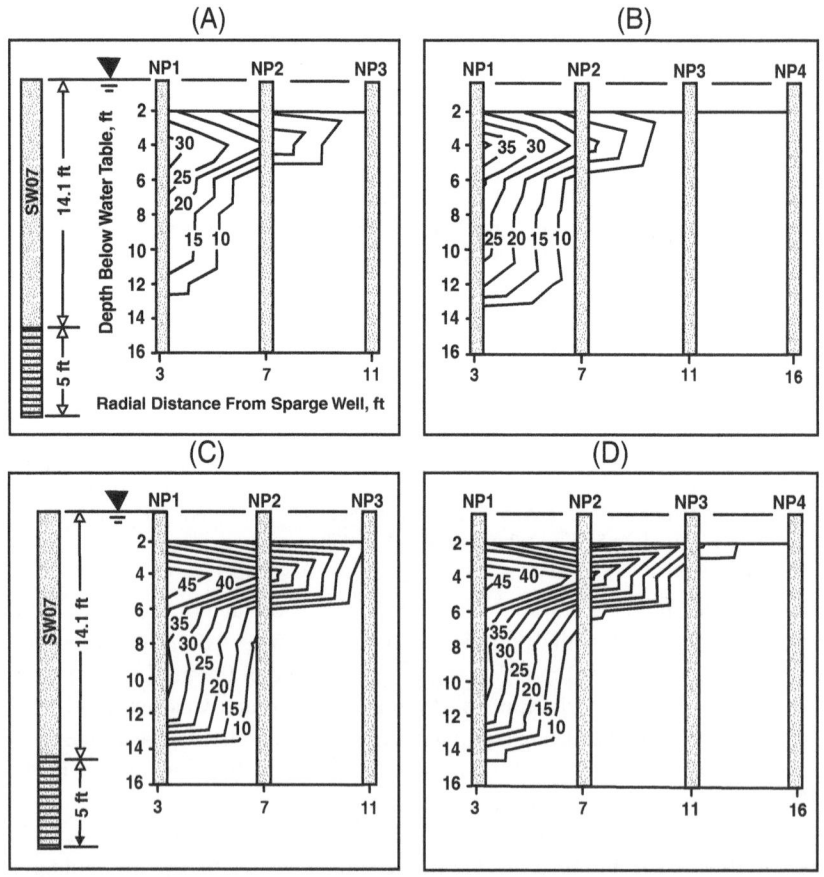

Figure 2.4 Air distribution profiles (percentage of pore volume containing air) during steady-state operation at sparge well SW07. (A) Air injection at 6.8 m³/min; (B) Air injection at 13.6 m³/min; (C) Air injection at 20.4 m³/min; (D) Air injection at 27.2 m³/min.

radial distance of 0.9 m from the sparge well increased slightly from 4 to 4.4 m
BWT when the flow rate was doubled from 13.6 to 27.2 m³/h.

Repeatability

Sparge well SW01 was operated at 19 m³/h for 0.5 hours before performing
measurements in access pipe NP12 (Figure 2.5). At 1.5 m from the sparge well,
the average measured value for air saturation was 36% between 0 and 2.4 m BWT.
The highest displacement, measured at 1.8 m BWT, was 46%. A second set of
readings was collected in the same borehole after 3.5 hours of operation and
revealed an average decline in air saturation of 18% between 0 and 2.4 m BWT.
After 4 hours of operation, the airflow was shut off. Air injection at the same flow
rate was resumed the following day, accompanied by moisture measurements at
the same locations and time intervals as the previous day. The second set of data
was almost identical to the first, yielding an average displacement of 36% between
0 and 2.4 m BWT, a high value of 46% at 1.8 m BWT, and an average decline of
20% after 3.5 hours of operation.

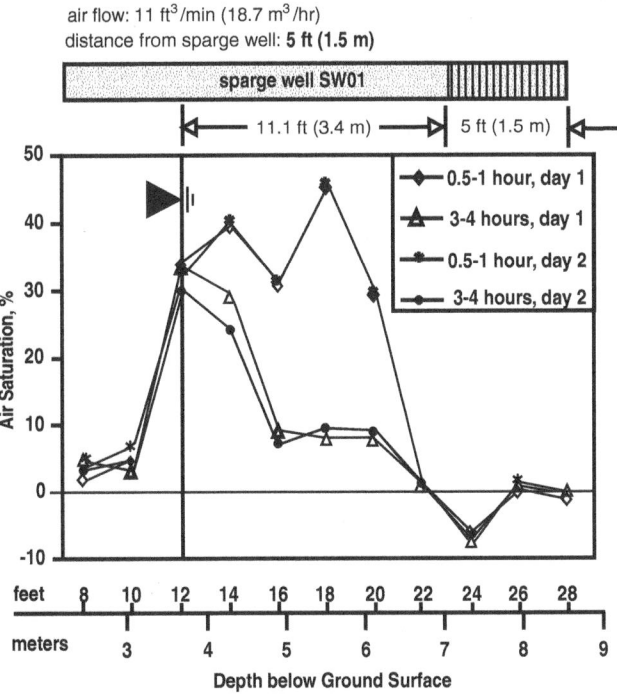

Figure 2.5 Percentage of air saturation during two injection cycles on separate consecutive
days. Airflow was shut off each day after 4 hours of operation.

DISCUSSION

Two phases of transience are reported to occur during IAS before steady-state air distribution is attained (Lundegard and Anderson, 1993). The first phase is an expansion of the air-saturated region controlled by the induced and hydrostatic pressure gradients. During this phase, mounding of the water table occurs due to displacement of groundwater by the initial surge of air. As the percentage of air saturation increases, the air-phase permeability of the surrounding soil also increases, establishing preferential flow near the sparge well where air saturation is highest. This preference is due to lower air-entry pressures that decrease exponentially with increased air content. As permeability near the well continues to increase, a point is reached where pore spaces in the outer limits become resaturated with water, possibly causing a decrease in the water table elevation as the displaced groundwater returns (Lundegard and LaBrecque, 1995).

During steady-state operation at a flow rate of 27.2 m^3/h, a maximum air saturation level of 52% was measured. However, if the intent is to acquire the largest expanse of air distribution, then cycled operation of the IAS system is needed to avoid steady-state consolidation. The desired frequency of a cycled operation will vary with site conditions and objectives. Since the consolidation phase does not occur uniformly, defining the point of peak expansion requires consideration of air saturation changes in both radial and vertical directions. In Figure 2.3A, the region of influence, as defined by the 10% air saturation contour, extended about 3.7 m horizontally near the water table and about 5.3 m vertically near the injection screen after 0.5 hours of operation. After an additional hour of operation, the contours shifted about 0.5 m laterally near the water table, while the lower limit of influence shifted upward about 0.8 m (Figure 2.3B).

Cycled operation has the added value of inducing groundwater movement away from the injection screen during startup and toward the injection screen following shutdown. This groundwater movement may mitigate oxygen diffusion limitations in regions of low air saturation by mixing pore water containing high and low levels of oxygen.

CONCLUSIONS

A dynamic network of air pathways was evident around two IAS wells. The general characteristics of air distribution consisted of a relatively rapid expansion phase followed by a consolidation phase. The latter began at the deepest elevations while radial expansion was still occurring at higher elevations. Steady state within the range of influence was attained within 12 hours at all measured flow rates. Air saturation levels reached greater than 50% during steady-state operation, but the best overall air distribution was attained during the transient phase, with measurable changes in water content occurring beyond 3.7 m from the sparge well. Increased air flow rates extended the depth of the air saturation profile, but the lateral limit of steady-state profiles remained essentially constant. Thus, the effect of greater air mass was to increase the bulk density of air distribution rather than expand the ROI.

Cycled air injection was capable of repeatable levels of air saturation during transience and steady-state operation. Cycling provided better overall air distribution by avoiding steady-state consolidation of air pathways and possibly mitigating diffusion limitations in regions of low air saturation. The homogeneous sands at this site required a pulsing frequency of only a few hours.

ACKNOWLEDGMENT

This work was funded by the Federal Aviation Administration–Alaskan Region under contract no. DTFA04-92-P-89250.

REFERENCES

Baehr, A.L. and M.F. Hult. Evaluation of Unsaturated Zone Air Permeability Through Pneumatic Tests. *Water Resources Research,* 27, 2605–2617, 1991.

Brown, R.A., R.J. Hicks, and P.M. Hicks. Use of Air Sparging for *In Situ* Bioremediation. *Air Sparging for Site Remediation*, Lewis Publishers, Boca Raton, Florida, 38–55, 1994.

Brown, R.A. and F. Jasiulewicz. Air Sparging: A New Model for Remediation. *Pollution Engineering,* 52–55, 1992.

Hinchee, R.E. Air Sparging State of the Art. *Air Sparging for Site Remediation,* Lewis Publishers, Boca Raton, Florida, 1–13, 1994.

Johnson, R.L., P.C. Johnson, D.B. McWhorter, R.E. Hinchee, and I. Goodman. An Overview of *In Situ* Air Sparging. *Groundwater Monitoring Review,* Fall, 127–135, 1993.

Kline, S.J. and F.A. McClintock. Describing Uncertainties in Single-Sample Experiments. *Mechanical Engineering,* pp. 3–8, 1953.

Lundegard, P. and D. LaBrecque. Air Sparging in a Sandy Aquifer (Florence, Oregon, U.S.A.): Actual and Apparent Radius of Influence. *Journal of Contaminant Hydrology,* 19, 1–27, 1995.

Lundegard, P.D. and G. Anderson. Numerical Simulation of Air Sparging Performance. *Proceedings of the 1993 Petroleum Hydrocarbon and Organic Chemicals in Groundwater: Prevention, Detection, and Restoration Conference,* 461–476, 1993.

Marley, M.C., D.J. Hazebrouck, and M.T. Walsh. The Application of *in Situ* Air Sparging as an Innovative Soils and Ground Water Remediation Technology. *Groundwater Monitoring Review,* Spring, 137–145, 1992.

Marley, M.C., E.X. Droste, and R.J. Cody. Mechanisms That Govern the Successful Application of Sparging Technologies. *87th Annual Meeting of the Air and Waste Management Association,* 1994.

Martin, L.M., R.J. Sarnelli, and M.T. Walsh. Pilot-Scale Evaluations of Groundwater Air Sparging: Site Specific Advantages and Limitations. *Proceedings of R&D 92 National Research and Development Conference on the Control of Hazardous Materials,* 318–327, 1992.

CHAPTER **3**

Subsurface Barrier and Recovery Trench for Contaminant Removal: Naval Arctic Research Laboratory Point Barrow, Alaska

James Baldwin, Ron Borrego, Amadeo Rossi, and Scott Horwitz

CONTENTS

SUMMARY

    This chapter addresses the selection, design, and implementation of a remedial system at the Naval Arctic Research Laboratory (NARL) in Barrow, Alaska (see Figure 3.1). The purpose of this system was to prevent existing, "active zone" groundwater contamination from migrating into the adjacent Imikpuk Lake, the source of local drinking water. A subsurface barrier (or "berm"), recovery trench, and treatment system comprised the remedial system.

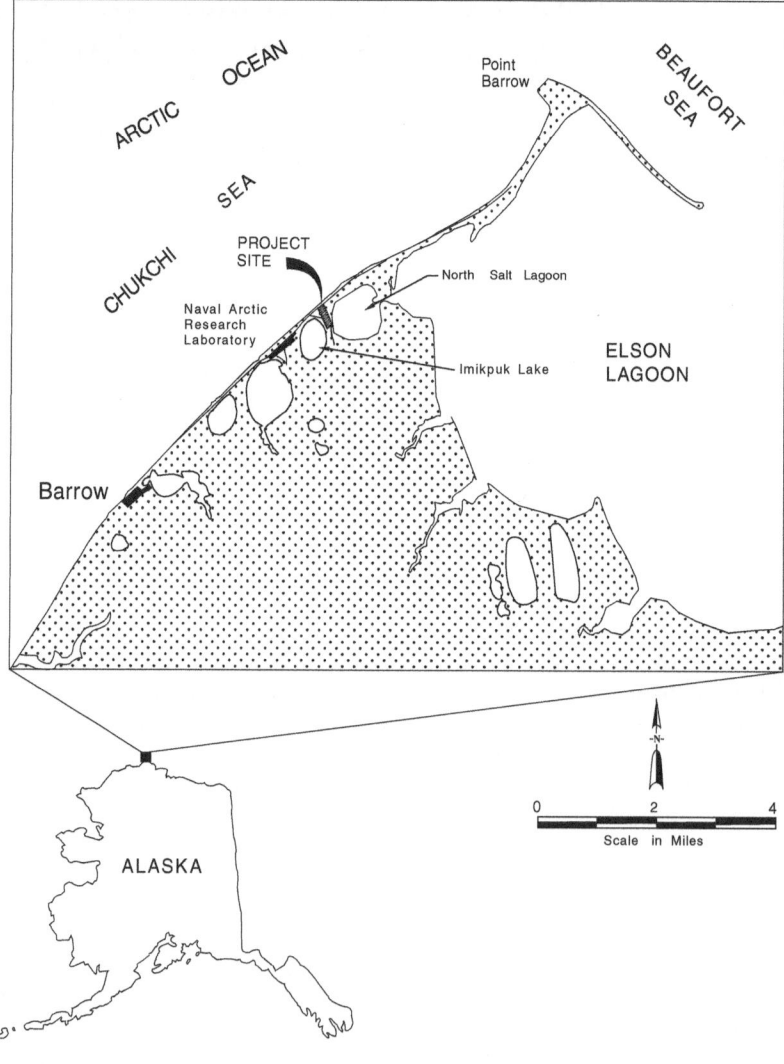

Figure 3.1 Location map.

SITE HISTORY AND BACKGROUND

The climate in Barrow is characterized by extremely cold winters and cool summers with constant winds averaging 4.5 m/s (10 mph). Temperatures range from –28°C (–19°F) in February to 4°C (40°F) in July. The region is relatively dry; generally less than 60 cm (2 ft) of snowpack accumulates during the winter.

In the area around NARL, the soil consists of unconsolidated marine and non-marine gravel, silt, and clay that extends from the surface to 15.2 m (50 ft) below the ground. Some finer deposits of silt, clay, and peat occur in drained lake basins and in places along beach ridges. The soil remains frozen at the surface through most of the year. During the summer, the surface layer of soil thaws slowly, typically reaching a depth of 1.8 to 3.4 m (6 to 11 ft). This thawed layer is termed the "active zone" (Figure 2.2). The active zone can be affected by the presence of heated buildings, the removal of the upper layer of insulating soil, snow cover depth, and saturation by saline active zone water or a mixture of hydrocarbons and water. Below the thawed layer is permafrost. Because the permafrost is impermeable to water, near-surface groundwater is confined to the active zone.

NARL is currently operated by the Ukpeaguik Inupiat Corporation (UIC), the local native village corporation. The primary environmental concern at NARL was contamination from four fuel spills associated with the airstrip. In August 1976, an underground pipe failure discharged an estimated 181,000 L (48,000 gal) of gasoline. In December 1978, approximately 94,500 L (25,000 gal) of JP-5 fuel and 1,058,400 L (280,000 gal) of gasoline were spilled in separate incidents. Approximately 529,000 L (140,000 gal) were recovered from the two spills, and the remainder was burned

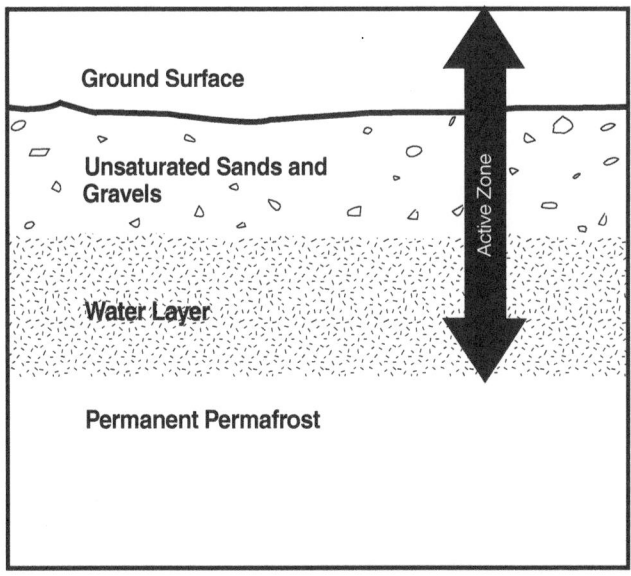

Figure 3.2 Cross-sectional detail defining the active zone.

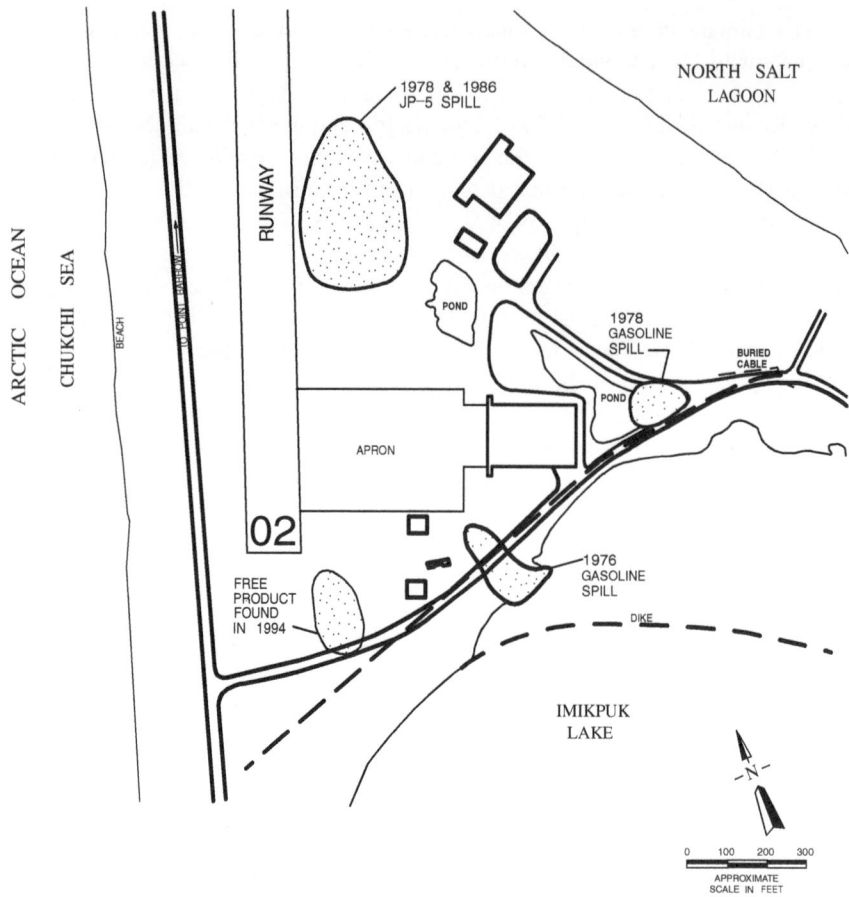

Figure 3.3 Airstrip fuel spill area, fuel spill locations.

off in July 1979. The fourth spill occurred in 1986 and involved an unknown quantity of JP-5 fuel. During the summer of 1994, free product was found in the active zone groundwater at the end of the runway. Figure 3.3 shows the approximate locations of these spills. As part of the land exchange agreement that relinquished ownership of the NARL facility to UIC, the U.S. Navy agreed to perform localized remediation of selected areas within the NARL.

PREVIOUS INVESTIGATIONS

Previous studies assessed the presence and extent of contamination at NARL (DOWL, 1986; NEESA, 1983; Reidel, 1988; SAIC, 1987, 1988, 1989a, 1989b,

1990). The studies included investigations of potential soil and groundwater contamination in the active zone and in surface water and sediments in Imikpuk Lake. Samples were generally analyzed for benzene, toluene, ethylbenzene, and xylenes (BTEX); total petroleum hydrocarbons (TPH); and occasionally lead.

Contaminant concentrations varied. In soil, a maximum of 467 ppm TPH was found in 1988 and 2600 ppm TPH in 1991. Concentrations in groundwater samples reached a maximum of 4.2 ppm gasoline-range petroleum hydrocarbons and 22.6 ppm total BTEX. Concentrations were highest in both soil and groundwater in locations adjacent to past fuel spills. In surface water samples from Imikpuk Lake, TPH was detected near the north shore in 1988 and lead was found in 1989. These findings indicated that the fuel contamination was migrating toward the lake.

REMEDIAL OPTIONS CONSIDERED

A feasibility study was conducted to determine the most economical and appropriate remedial action for the site. At the time, only soil and groundwater contamination was suspected. After an initial screening, four potential options were identified:

1. No action
2. Containment/treatment
3. On-site incineration
4. Off-site landfill

The first option (no-action) did not protect human health and the environment because it would not provide treatment or containment of the source area contamination. Petroleum contaminants would likely migrate to the lake. For this reason, the no-action option was dismissed as unacceptable.

The second option (containment treatment) consisted of physical barriers to prevent the contamination from entering the lake, thereby protecting human health and the environment. While this option addressed the exposure pathway and potential contaminant mobility concerns, it would not reduce the volume of contaminated soil on site. Thus, this option would require implementing institutional controls, such as limitations on excavation and additional monitoring to ensure its long-term effectiveness.

The third option (on-site incineration) involved destroying petroleum contaminants bound in the soil. Incineration is a proven technology with a high degree of effectiveness. However, the cost to incinerate saturated soil at a remote site such as Barrow was considered prohibitive.

The fourth option (off-site landfill) involved disposing of all contaminated soil at an offsite landfill. This action would protect human health by removing source area contamination. Although the mobility of the contaminants would be controlled by placement in a landfill cell, the toxicity and volume of the contaminated soil would remain undiminished. This option was rejected because it did not address contaminated groundwater. Also, because of the site's remoteness, the cost of removing contaminated soil to an off-site landfill would have been prohibitive.

SELECTED OPTION

After evaluation, the second option was determined to be the most feasible. In this option, a subsurface containment berm and recovery trench, specifically designed for the Arctic region, would be installed. Groundwater would be extracted upgradient of the containment berm and then treated. This option was the most economical and prevented the migration of contaminants into Imikpuk Lake.

This containment berm design uses two barriers to control groundwater movement. The primary barrier is a permafrost ridge created by installing rigid insulation near the ground surface. The insulation slows the downward transfer of heat under the insulation during the summer months. The secondary barrier is a vertical geomembrane curtain wall embedded in the center of the permafrost ridge. This barrier provides backup protection should thermal conditions threaten the integrity of the permafrost ridge. To prevent the ponding of melted water behind the ridge, a recovery trench was placed upgradient to collect water flowing toward the permafrost ridge. This recovery trench collects and delivers water to the active zone water treatment plant.

DESIGN CONSIDERATIONS

A primary concern in designing the containment berm was the depth and characteristics of the permafrost zone. Drill samples were taken along the proposed route of the project at the time of maximum thaw (summer 1995), and it was determined that the active zone extended to a depth of 1.6 m (5 ft) along the containment berm alignment. Based on this information, the containment berm was designed to extend from 0.8 m (2.5 ft) below grade to 3.2 m (10.5 ft) below grade, about twice the depth of thaw in an average year. This ensures that if the permafrost barrier erodes, the geomembrane will continue to contain the groundwater. The top 0.8 m (2.5 ft) of the berm contains rigid insulation lying immediately above the geomembrane (the insulation was also covered with plywood for protection), which was filled to grade with roadway bed material. This material was necessary to protect the rigid insulation and trench work from vehicular traffic.

The use of rigid insulation to control thaw depth in Arctic zones is well established. Applications include using rigid insulation to elevate the permafrost, providing stability to foundations and trench work. During the design phase of this project, thermal analysis was used to determine an optimum configuration of insulation size and placement. A minimum distance between the permafrost berm and the recovery trench was also determined to ensure that the presence of flowing water in the trench did not thermally erode the berm.

Contaminated groundwater intercepted by the recovery trench is pumped to the treatment plant built specifically for this remedial action. The treated groundwater is discharged to an existing sewage lagoon for final disposition.

The recovery trench was designed to draw away groundwater and prevent it from flowing over the containment berm by having a minimum allowable depth of 0.8 m (2.5 ft). This depth placed the trench below the top of the berm, but it would not extend to the permafrost zone. Consequently, the recovery trench was designed to extend from grade down to 1.2 m (4 ft) below grade, which is below the top of the containment berm yet above the surface of the permafrost zone during periods of typical maximum thaw.

The location and alignment of the containment berm and recovery trench (the project alignment) extends for a length of 518 m (1700 ft), roughly paralleling the shoreline of Imikpuk Lake. Approximately 366 m (1200 ft) of the total length lies along an existing roadway, with the remaining 152 m (500 ft) following the shoreline where the roadway curves away from the lake. Figure 3.4 shows a typical view of the right-of-way. Several factors were involved in choosing the project alignment. Primary among these was that the containment berm serves as a barrier to prevent hydrocarbon contaminants within the active zone groundwater from migrating into Imikpuk Lake. Therefore, the containment berm must lie between the areas of contamination and the lake. Other design considerations were

- The existing roadway base and adjacent fine- to coarse-grained gravels provided a stable and uniform support to anchor the geomembrane liner into the permafrost.
- Soil investigations along the project alignment indicated that the soil was relatively free of saline zones or ice wedges that could thaw, leading to potential migration under the geomembrane liner.
- There were no thaw regions along the project alignment that would compromise the permafrost berm.

Surface water from areas of the runway apron collects along the road. In some locations, the water flows over the top of the roadway into Imikpuk Lake. Surface

Figure 3.4 Typical view of the project right-of-way.

water also collects in a small backwater pond adjacent to and east of the largest hangar. Unless the surface water is redirected, most of it would end up in the recovery trench and have to be treated. A drainage swale was therefore constructed to divert the surface water away from the recovery trench.

THERMAL MODELING

Thermal modeling was performed to more accurately understand the thermal interactions between the permafrost ridge, recovery trench system, rigid insulation, and the seasonal fluctuation of air temperature above the system. This modeling predicted the depth of thaw for a set of climatic conditions and estimated the effectiveness of insulating layers in creating the permafrost ridge.

The thermal modeling used for the NARL site was dependent on two factors. The first factor was the seasonal advance and retreat of the active zone within the permafrost. As spring temperatures and sunlight hours increase, the depth of thaw or active zone increases from the ground surface down. The rate and depth of advance are dependent on the type of soil, with native tundra silt experiencing less active zone development than beach sands or artificially placed gravel. The second factor was the thermal erosion created by meltwater. The recovery trench was designed to intercept meltwater and provide a source of thermal energy to the system. The modeling was used to determine a recovery trench location that did not affect the thermal integrity of the containment berm.

Estimates of Active Zone Groundwater Flow

To determine the thermal impact of the recovery trench on the containment berm system, the maximum volume of water collected by the 518-m- (1700-ft-) long trench was estimated based on established trench dewatering formulas and empirical values of hydraulic conductivity for the predominant beach sand material at the site. It was assumed that the flow of water into the trench was limited by the hydraulic conductivity of the soil; that is, any precipitation ponding on the ground surface would percolate into the groundwater as the soil allows.

Modeling Simulation of the Active Zone

Seasonal thaw depths of the containment berm and recovery trench system were predicted using TQUEST, a proprietary finite-element geothermal simulation model developed by Northern Engineering and Scientific, Anchorage, Alaska (Northern Engineering, 1995). TQUEST simulated the development of the active zone, taking into account climatic data, thermal input from the active zone groundwater, and soil type.

Long-term climate data collected for Barrow by the National Oceanic and Atmospheric Administration were used in the modeling simulations. Ground temperatures needed to initialize the finite-element model grid were taken from long-term simulations of a "typical" 1.5-m- (5-ft-) thick gravel pad in Barrow, underlain with icy sandy silt and silty sand. The simulations assumed yearly formation of a snowdrift along the sideslope of the gravel pad. The snowdrift acts as an insulator, limiting beneficial soil cooling along the toe-of-slope during the winter, and contributing to increased thaw depth in the summer. Three years of climatic data were included for the thermal modeling simulations.

Water flow in the recovery trench was modeled as a migrating vertical thermal flux boundary, simulating downward thermal erosion of permafrost by water flow. Water was assumed to enter the trench at 5°C (40°F) and cool to approximately 0°C (32°F) while en route to sumps located along the recovery trench. Three sumps were assumed, each receiving a third of the total flow.

Rigid insulation (3.4 m [8 ft] wide by 10 cm [4 in.] thick) was centered over the berm, resulting in a composite R-value of 24 for the system. The centerline separation between the containment berm and the recovery trench was modeled at 4.8 m (16 ft) to fit within the roadway footprint. The simulations also considered the use of insulation in the recovery trench to concentrate thermal erosion and provide a passive-thawing mechanism to open the trench at the beginning of each thaw season. In the final design, this approach was abandoned in favor of heat trace to thaw the trench in spring.

The results of the simulations indicated that a raised berm of frozen material would be created and maintained by insulation laid horizontally near the ground surface, even during the periods of maximum active zone development. The 4.8-m (16-ft) separation between the containment berm and recovery trench was found adequate to protect the berm from the thermal effect of the recovery trench.

SYSTEM FEATURES

Components of the selected remedial action system include the following: containment berm, recovery trench, drain collection sump, drainage swale, and water treatment system. This remedial system was completed in June 1996 and is currently in operation and being monitored for performance capability.

Containment Berm

The containment berm consists of a permanent impervious geomembrane installed vertically into the permafrost (Figure 3.5). The geomembrane was installed in a trench that was 0.3 m (1 ft) wide and 2.4 m (8 ft) deep, running the length of the project alignment. A geotextile cushion was used to pad the membrane, preventing possible damage by the trench wall. Following placement of the membrane, the trench was backfilled with a sand slurry and allowed to freeze. Rigid insulation was centered on top of the trench, over which sheets of marine-grade plywood and approximately 0.6 m (2 ft) of protective gravel were placed. The gravel became the new surface of the roadway in those locations where the berm is installed in the road. The insulation was installed during the winter, trapping the lowest subgrade temperatures possible. The insulation minimizes thawing of the permafrost beneath it, thus protecting the integrity of the containment berm.

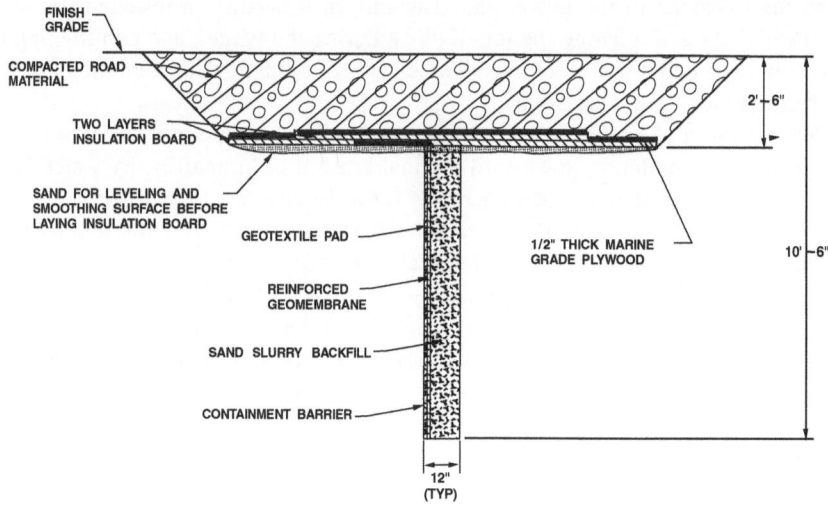

Figure 3.5 Containment barrier detail.

Piezometers were installed upgradient and downgradient of the containment berm to measure active zone water levels. A thermistor string also was installed along the containment berm to measure soil temperatures, indicating the depth of permafrost. This information enables monitoring of the containment berm's performance and gives early warning of a breach by thermal or hydraulic erosion. If it is determined that the thaw depth of the permafrost could threaten the integrity of the containment berm, passive thermal radiators can be installed to reestablish the berm.

The containment berm requires little or no direct maintenance. However, it is important to keep the roadway in good condition, with a minimum of 0.6 m (2 ft) of bed material, to protect the containment berm.

Recovery Trench

The recovery trench was installed upgradient and parallel to the entire length of the containment berm to intercept active zone groundwater and prevent its ponding against the barrier. There was a minimum separation of 4.8 m (16 ft) between the recovery trench and the containment berm, so that heat from the recovery trench system and groundwater flowing in the recovery trench did not thermally erode the containment berm (Figure 3.6).

The recovery trench was approximately 10.3 m (1 ft) wide and 1.2 m (4 ft) deep, and contained two slotted high-density polyethylene (HDPE) drain pipes and highly permeable gravel. The pipes and gravel transfer contaminated active zone water to drain sumps for pumping into the groundwater treatment system. The recovery trench was thawed at the beginning of each summer to capture active zone groundwater in the trench from the outset of the spring melt. The two capture pipes had heat trace

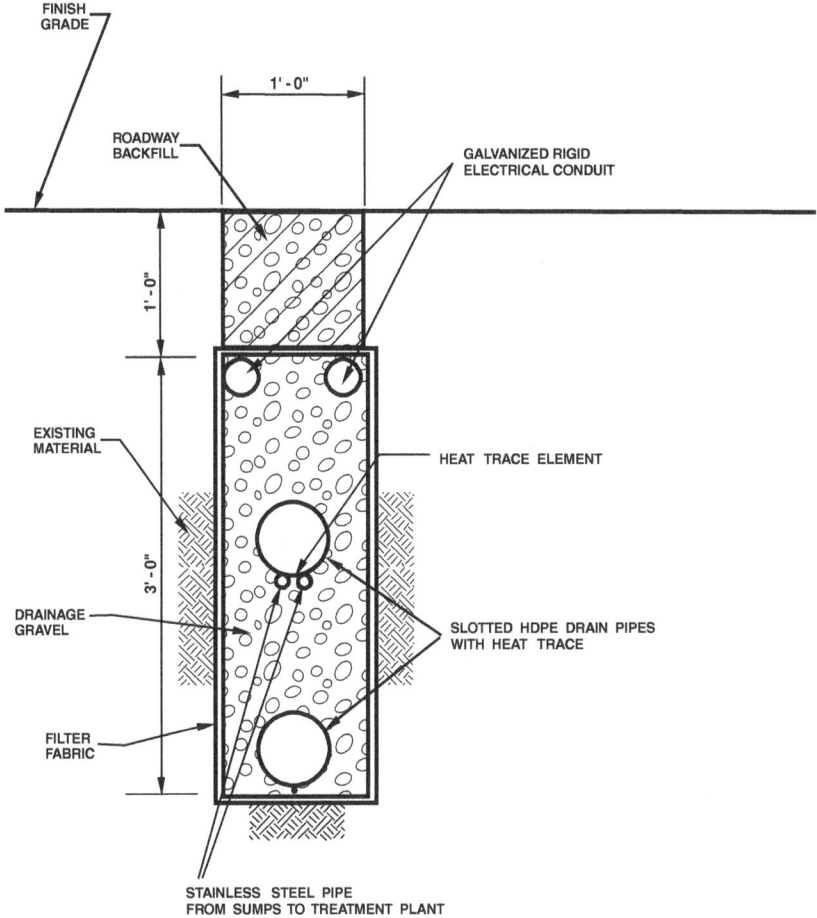

Figure 3.6 Recovery trench detail.

elements that are controlled separately; this allowed the upper pipe to thaw first, and the lower pipe to thaw as the seasonal frost thaw migrated downward.

Drain Collection Sump

Drain sumps were spaced along the recovery trench at three locations to collect contaminated groundwater for pumping into the treatment system. Sumps were installed to a depth of approximately 2.4 m (8 ft) below the bottom of the trench to provide a collection reservoir for the pumps. Drain sumps consisted of insulated, 60-cm- (24-in.-) diameter HDPE pipes accessible at the surface for the convenient installation and removal of the pumps. A geotextile filter fabric was used to prevent silt and fine sand particles from entering the drain and sumps, but the geotextile fabric did not extend into the roadway (see Figure 3.7).

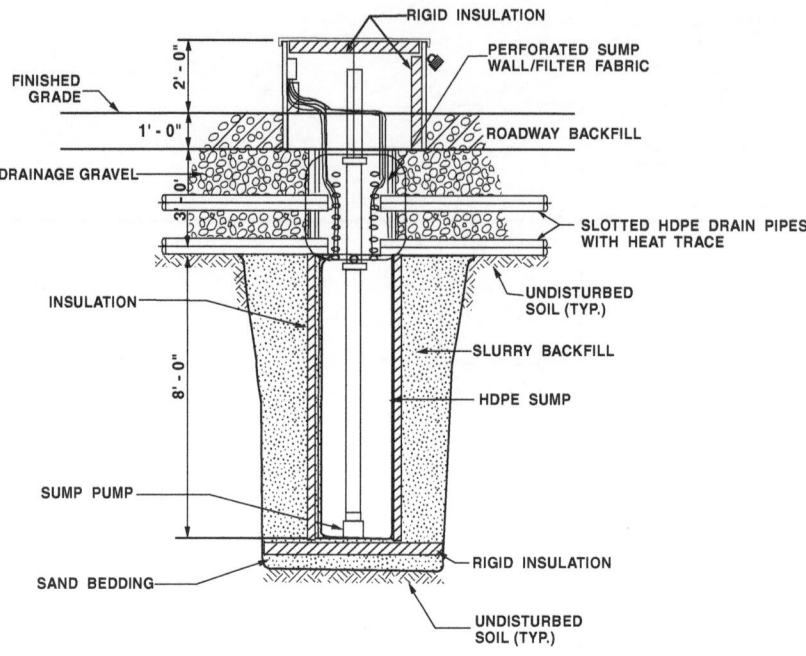

Figure 3.7 Groundwater collection sump detail.

Electric heat elements were installed to thaw the sumps in the spring prior to starting the system.

Water Treatment System

Active zone groundwater captured by the recovery trench was treated prior to discharge to an existing sewage lagoon. The treatment facility consisted of an oil-water separator (OWS), hydrocarbon plate stripper, and granular activated carbon (GAC) filters. The OWS removed floating product and reduces diesel and gasoline concentrations to approximately 100 ppm. From the OWS, the water gravity fed to a 1040-L (275-gal) batch tank; from there the groundwater was pumped to the hydrocarbon plate stripper, which further reduced the concentrations to less than 10 ppm. From the hydrocarbon plate stripper, the water was pumped to the GAC system for final organic removal. The GAC system included two GAC units in series. Following treatment and prior to release to the sewage lagoon, the treated ground-water was collected in 38,000-L (10,000-gal) tanks for verification sampling.

METHOD OF CONSTRUCTION

The remedial action system was constructed in phases, with each phase scheduled to take advantage of the seasonal conditions at the site. Phase I, completed during

Figure 3.8 View of the treatment plant building and thermal radiators.

the 1995 summer construction season, consisted of constructing the water treatment plant building and site preparatory work for the installation of the containment berm and recovery trench. A view of the water treatment plant building, including the thermal siphons, is shown in Figure 3.8. The site preparatory work included surveying, locating and relocating utilities, initial excavations, and construction of the treatment plant building. In addition to the on-site work, barge shipment of all materials occurred during this phase to avoid the higher costs of air freight during the winter months.

Phase II, which was performed during February through April 1996, included the construction of the containment berm and recovery trench. A large rock saw (see Figure 3.9) was used to cut the trench for the vertical barrier (see Figure 3.10). The completed liner with geotextile backing is shown in Figure 3.11 with sand slurry placement in Figure 3.12. The placement of the road backfill over the insulation board and plywood is shown in Figure 3.13.

Performing the work during the cold season helped to freeze the sand slurry and ensure that the coldest possible temperature was trapped beneath the insulation covering the containment berm, encouraging the formation of an elevated permafrost barrier. During this phase of construction, free-flowing product was encountered during the excavation of the trench for the containment berm, which required the realignment of the initial 2113 m (700 ft) of the project closer to the lake, and the installation of two additional drain sumps to recover the free-flowing product.

Phase III consisted of the installation and start-up of the treatment plant. This work began with the construction of each unit on a portable skid, which allowed for easy transport to the site. This work was completed during May, and start-up and testing occurred during the first weeks of June 1996.

Figure 3.9 Large rock saw for construction of the containment berm.

Figure 3.10 Trench excavation using the rock saw.

Figure 3.11 Completed trench ready for slurry.

Figure 3.12 Placing slurry behind the filter fabric and geotextile.

Figure 3.13 Placing backfill over the insulation board and plywood to create the road section.

Phase IV of the construction, which was completed in early July 1996, included road filling and compaction after the active zone thawed.

Phase V of the project involved the operation and maintenance of the treatment plant from July through September 1996. In September, however, the system was still producing active zone groundwater. Because of this, the operation and maintenance phase was continued through December 1996.

RESULTS OF FIRST YEAR OF OPERATION

During the first season of operation (June through December 1996) of the subsurface containment berm, recovery trench, and treatment system, approximately 2,725,000 L (720,000 gal) of active zone groundwater was treated and discharged to the local sewage lagoon. Approximately 204,000 L (54,000 gal) of free product was recovered from the collection system. The plant operated 7 days a week, 12 hours a day. Product recovery from the collection system reached 50% free-flowing product and 50% water during the months of November and December. The high fuel recovery rate for a 1-day period was 55,514 L (14,667 gal) of free product during November.

REFERENCES

DOWL Engineers. Fuel Spill Assessment Study. Prepared for Ukpeaguik Inupiat Corporation, August 7, 1986.

NEESA (Naval Energy and Environmental Support Activity). Initial Assessment Study of Naval Arctic Research Laboratory (NARL), Point Barrow, Alaska. NEESA 130-026, 1983.

Reidel Environmental Services. Naval Arctic Research Laboratory (NARL), Point Barrow, Alaska. Prepared for Oil and Hazardous Material Abatement ESSM Base. 15 December 1988.

SAIC (Science Applications International Corporations). Evaluation of Hazardous Materials and Potential Environmental Contamination at the Naval Arctic Research Laboratory, Point Barrow, Alaska. Prepared for the U.S. Department of Energy, November 1987.

SAIC (Science Applications International Corporations). Evaluation of Potential Contamination at Naval Arctic Research Laboratory, Point Barrow, Alaska. November 1988.

SAIC (Science Applications International Corporations). Final Evaluation of Site Options at NARL, Point Barrow, Alaska, July 1989(a).

SAIC (Science Applications International Corporations). Investigation of Environmental Concerns at the Naval Arctic Research Laboratory, Point Barrow, Alaska. Prepared for the U.S. Department of Energy. Seattle, Washington, April 14, 1989(b).

SAIC (Science Applications International Corporations). (1990). Evaluation of Potential Environmental Contamination at the Naval Arctic Research Laboratory, Point Barrow, Alaska. Prepared for the U.S. Department of Energy. Seattle, Washington, April 20, 1990.

Tryck, Nyman, and Hayes (TNH). Naval Arctic Research Laboratory, Fuel Spill Investigation. Prepared for the U.S. Navy Western Division Naval Facilities Engineering Command. Anchorage, Alaska, December 1987.

URS Consultants, Inc. Draft Supplemental Site Characterization, Airstrip Fuel Spill Area, Imikpuk Lake, North Salt Lagoon, and Selected Meltwater Ponds, Naval Arctic Research Laboratory, Point Barrow, Alaska, December 6, 1993. Prepared for U.S. Navy CLEAN, N62474-89-D-9295. Seattle, Washington, 1991.

CHAPTER 4

Strategies for Development of Cost-Effective Amelioration Procedures for Oil Spills in Cold Regions

Peter J. Williams, W. Gareth Rees, Dan W. Riseborough, and T. Les White

CONTENTS

INTRODUCTION: THE GROUND CONTAMINATION PROBLEM

The cold regions of the world are usually remote and sparsely populated, and consequently it comes as a surprise to some that extensive problems of pollution occur there. In part these problems have arisen precisely because the sites are distant and little known. Yet the detrimental effects have proven to be widespread, with pollution of water supplies and the entry of noxious substances into food chains, and indeed destruction of the environment to an extent (in, for example, the recent Russian oil spills) as to greatly disturb local communities. This chapter considers the problem of oil spills on the ground surface (rather than directly into a water body), and the conclusions, moreover, apply broadly to many other kinds of ground contamination. Because much of the contamination in rivers and oceans originates with spills onto the ground surface, the topics have an extensive relevance. Some of the studies proposed also have application to spills in warmer regions where there is nevertheless a substantial winter frozen layer.

Responses to spills in remote cold regions fall into several categories: containment of the pollutant (Boitnott et al., 1997, Dash et al., 1997, Iskandar and Sayles, 1997), removal of contaminated material and possibly its incineration, facilitated chemical or biodegradation and, sometimes, doing nothing. In the cold regions these responses tend to be particularly difficult and costly, sometimes enormously so. More sophisticated procedures involving removal of contaminated soil, its "cleansing" by chemical, mechanical, thermal, and other procedures using special equipment, although widely used in urban environments, are usually impractical in remote locations where spills are extensive. In spite of a rapidly growing scientific and geotechnical literature (Virtual Conference, 1998), it is clear that research and development of appropriate remedial procedures are much needed. All too often, response is based on an assessment of the extent of the contamination and the longer-term socioeconomic implications without a sufficiently differentiated and considered approach to the ameliorative procedure — simply because the knowledge for this is not available. This chapter (developed from a preliminary study by Williams et al., 1996) argues that by using existing knowledge together with further research into the behavior of freezing ground, prediction of the longer-term behavior of the spilled oil will be possible on a site-specific basis with much greater accuracy and detail. Such prediction will allow a far more sophisticated site-specific assessment of the appropriate and the most cost-effective response procedure. Because of the enormous expenditures potentially involved, we see this as research offering an unusually high pay-back.

Even though the ground-surface layer is the first active medium spilled oil meets before entering the groundwater system, and one which exhibits very complex behavior, a recent international bibliographic search showed that very little is available in the literature on the interaction of oil and other hydrocarbons with freezing soils (Geotechnical Science Laboratories, 1994, but see also Virtual Conference, 1998). Much more is known about oil in unfrozen soils.

NEW STRATEGIES BASED ON PREDICTION OF SPILL EFFECTS

The choice of long-term response to an oil (or other) spill is most appropriately made when there is a good prognosis of what will happen, with and without a response. Surprisingly little consideration has been given to predicting the effects over time of the spill. For example, in remote northern locations, if it can be established with some certainty that the contaminant will not move from its existing location by virtue of the conditions there, then it may well be defensible to do nothing — a very cost-effective conclusion. On the other hand, if it is established that toxic substances will with time move, such as, for example, to pollute drinking water, or to threaten fisheries (by entering a river), then a geotechnical response is necessary. This response might be to contain the contaminant and immediately the questions of defining the appropriate barrier and its location arise, which can only be resolved if there is a full understanding of the processes set in motion by the spill. The sooner the processes specific to the site are resolved, the more rapidly the appropriate response can be decided upon.

This chapter considers the changes that the contaminant produces in the soil at the microscopic level. Then it is shown that such changes fundamentally alter a wide range of soil properties, which in the freezing conditions lead to specific changes in bulk properties and thus in the form and behavior of the ground surface layers and of the contaminant itself. Modifications of the surface occur which may be detectable by remote sensing, the significance of which will be enhanced by interpretative procedures based on the understanding of the soil behavior at the microscopic level.

It will be demonstrated that by integrating the results of research into these various aspects, a logical and systematic approach to the problem of cost-effective response can be developed. Several apparently separate research strategies are combined to give an overall strategic approach to the oil spill problem.

THE GROUND IN COLD REGIONS

The thermodynamic properties of freezing soils and the effects of phase change dominate ground behavior and demand special consideration.

Characteristics of Soils at Freezing Temperatures

The special properties of soils at freezing temperatures arise because of the particular nature of the freezing/thawing process within a fine-porous system (Williams and Smith, 1991). Most fundamental and consequential is the modification of the freezing point of the water. Because of capillarity and particle surface adsorption forces, highly significant amounts of the water in soils remain unfrozen at temperatures down to several degrees below 0°C. The amounts are dependent on the lithology and grain-size composition, and temperature, of the soil (Figure 4.1). The depth of the active layer and, indeed, the exact distribution and extent of permafrost depends on the thermal conductivity and heat capacity of the soil, which are highly

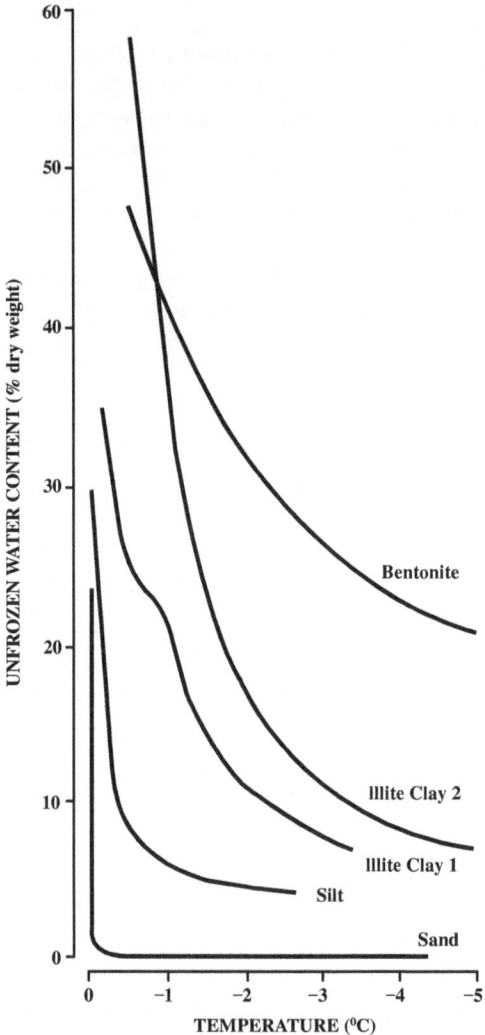

Figure 4.1 Unfrozen water content vs. temperature for a number of soils. From Williams and Smith, 1991.

dependent on the relations shown in Figure 4.1. The thermal conductivity is modified as the proportions of ice and water change, while the heat capacity of freezing soils is primarily a matter of the latent heat of fusion, exchanged as the proportions of water and ice change with changing temperature.

Particular pressure and stress conditions are associated with the range of freezing "points." The thermodynamic potential of the water changes as the temperature falls (effectively, a "suction" or fall in pressure occurs), so that water moves along temperature gradients. If water accumulates it gives rise to frost heave — a phenomenon well known to engineers. At the same time the ice phase exerts an expansive pressure. Before these effects become large enough to affect engineered structures, the soil has undergone changes in its internal, microscopic structure (Chamberlain and Gow, 1979). The thermodynamic effects have equally fundamental consequences for the strength and creep properties of frozen ground, which have been elucidated only recently (e.g., Ladanyi and Shen, 1989; Fremond and Mikkola, 1993).

Climatic Controls

The depth of the active layer (which freezes and thaws annually) and the distribution of permafrost are highly dependent on the heat exchange between the ground surface and the atmosphere. This heat exchange is determined by the physical nature of the surface layer (the vegetation, the surface of the soil and the microrelief and topographic exposure — all local site specific conditions) and its radiative, conductive, and evaporative properties. A further important factor is the thermal properties of the ground material (soil) itself. The stage is set by the general climatic conditions of the region.

This chapter cannot deal with all these matters in detail. They lie at the heart of the study of geocryology (Andersland and Anderson, 1978; Johnston, 1981; Williams and Smith, 1991). There is ample evidence in the terrain in the cold regions of the importance of all these effects in shaping the form and behavior of the near-surface layers of the ground (Washburn, 1978).

IMPORTANCE OF GROUND CONDITIONS FOR THE FATE OF THE SPILLED OIL

When oil is released on the ground surface it will begin to move into the ground and there may also be runoff over the surface. The proportions will depend, of course, on the rate of release and on the nature of the ground. There is a range of conditions that may be met, and the rate at which penetration into the ground occurs is very important. The subsequent movement within the ground and the nature of the interaction between the oil and soil constituents are of fundamental importance in defining the appropriate response.

The general characteristics of the ground upon which the oil falls are obviously important in determining the fate of the oil (Solntseva and Gruseva, 1997). But even more important are the degree and manner in which the oil itself modifies the soil properties and behavior, to the extent of such modifications essentially defining the subsequent developments in the contaminated ground. Thus we need to know more about the sequence and nature of changes that may take place in soils following addition of oil before we can predict the long-term fate of the oil at a particular spill site. In other words, while we now substantially understand the special properties and behavior of freezing ground in general, we do not know how these are modified by the addition of oil. In the following sections, the need for investigation of a number of properties is examined.

Thermodynamic Properties of the Ground Surface

Oil covering the ground surface so changes the properties relevant for energy exchange, radiational, convective/conductive and evaporative/condensative, that, inevitably, the energy balance will be substantially disturbed. Generally this will lead to warming of the ground and a deepening of the active layer, although in certain cases the opposite might apply. In either case there is a further, progressive disturbance of the ground surface. Where the regional mean ground temperatures are, prior to the spill, only −1°C or so, extensive thawing of permafrost may occur, in ice-rich ground, with major subsidences, and substantial surface instability. It is thus particularly important that the magnitude of the effects of the modification of the ground surface energy exchange on ground temperatures be more fully established.

Thermal Properties of the Soil

The addition of oil to the soil will, in many cases, cause sharp changes in the values of the soil thermal properties, not merely because of the presence of the oil (with its own specific values of thermal properties) but mainly because of the effects on the freezing process and the modifications in the exchanges of latent heat. The changes in thermal properties are likely to significantly modify the depths of freezing/thawing with similar implications for ground surface disturbance as noted above.

Thermodynamic Properties of the Soil

When an additional component is added to a soil, the physical-chemical conditions in the soil are changed. Exactly what these changes will be depends on the nature of a pollutant. But, with a freezing soil the changes will normally evidence themselves in a change in the unfrozen water content vs. temperature relation (Figure 4.1) reflecting a modification of the freezing point relations. In the case of oil one can surmise that the aggregation of the particles and the behavior of particle surfaces will be modified, that is, the capillary and adsorption phenomena known to control the proportions of unfrozen water and ice are modified. Associated with this are modifications of the energy status of the water and of the mechanical states within the soil. The permeability of the soil to unfrozen water (with its contained substances) will be changed. These changes will find expression in some accumulation or loss of ice, that is, in a renewed, continuing frost heave or a volume decrease of the frozen material, with extensive subsidence possible if total thaw occurs. These effects also lead, in turn, to additional modifications of the thermal properties and ground thermal regime. Thus while originating at the level of the soil pores and particles, the effects can soon become apparent at the ground surface where there may be minor or major changes in the form and elevation of the surface in a locally highly variable manner.

Soil Microstructure

The thermodynamic behavior of a freezing soil is expressed in the microstructure of the soil (see e.g., Yershov, 1990; Greshichev et al., 1992; and White, 1995). White and Williams (1994) showed, in microscopic thin sections of four soils, major changes in soil aggregation and porosity (including development of new pore shapes) as a result of freeze–thaw cycling. Figure 4.2A shows the fabric of a silt soil as it appeared following a preliminary compaction. Figure 4.2B shows the same material after being subjected to a cycle of freezing (to a temperature of approx. –2°C) and thawing. Figures 4.2C and 4.2D show the effects of three and ten freeze–thaw cycles. Elongated pores are the sites of ice segregations, and particle aggregation is also modified. Of particular interest is the progressive nature of the structural modifications with successive freeze–thaw cycles. Examination of similar structures in five types of soil showed that the progressive changes gradually weaken as the number of cycles increases (seen also in comparing Figures 4.2C and 4.2D).

Such modifications of fabric cause changes in many soil properties (White, 1995). The thermal, mechanical, and hydraulic properties will all be modified in varying degrees. The susceptibility to frost heave has been observed (Geotechnical Science Laboratories, 1993; 1995) to increase through successive freeze–thaw cycles.

The microstructural changes are to be related to stresses developed within the freezing soil that are effective in a highly differentiated fashion at the microscopic level. An early attempt made by Miller (1978) to analyze these stresses is shown in Figure 4.3A. More recently Ershov (1999) has produced the analysis shown in Figure 4.3B. It is not possible to describe these models in detail in this chapter. We believe that the redistribution of pore space and the modification of particle aggregations are ascribable to stress differentiation of this kind. In addition, as Yershov (1990) explains, they are also responsible for the development of fissures and other macroscopic discontinuities larger than the structures shown in the figures but small enough to be often overlooked in considering the mechanical and fluid conductive/interactive properties.

The movement of a fluid contaminant, the permeability of the soil to the contaminant, must relate closely to the soil structure and microstructure. Equally important, the surface area of soil particles and aggregates and the arrangement of void space defines the environment for the action of biological (or chemical) agents in hydrocarbon breakdown.

We are only now initiating studies of the microstructures of oil-polluted soils. But the origin of microstructures in stresses arising from thermodynamic behavior of the soil/ice/water system illustrates a most important hypothesis: the addition of oil will itself lead to a series of microstructural changes, following from the thermodynamic effects outlined above, continuing through successive freeze–thaw cycles. Just as the newly compacted soil showed a very marked reaction to the initial freeze–thaw cycling, with a gradually declining degree of alteration in further freeze–thaw cycles, so will the material newly and fundamentally modified by the addition of oil be susceptible to initially large and then declining microstructural changes with freeze–thaw cycling. Similarly, with time and if there is no further

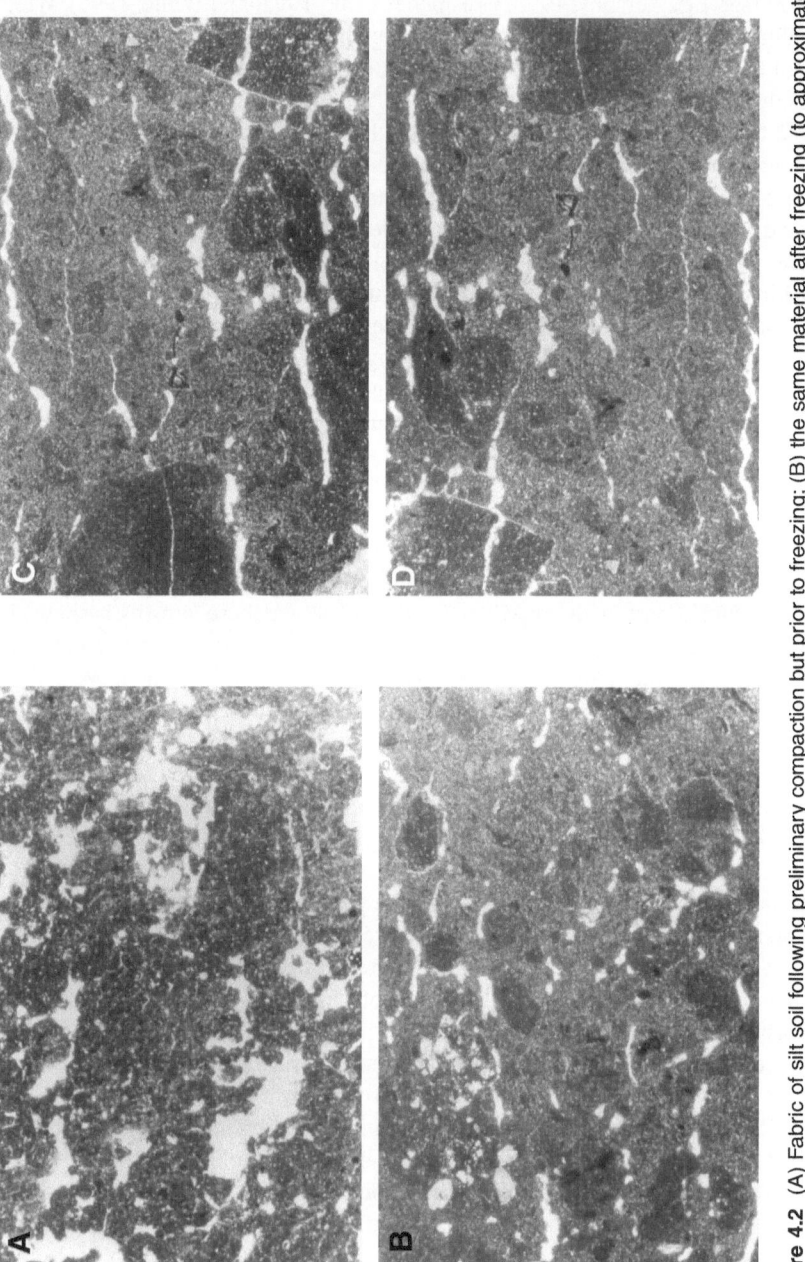

Figure 4.2 (A) Fabric of silt soil following preliminary compaction but prior to freezing; (B) the same material after freezing (to approximately −2°C) and thawing. Each frame is 13.5 mm long. (C) Similar material as in 4.2B but after three freezing-thaw cycles. (D) After ten freeze–thaw cycles.

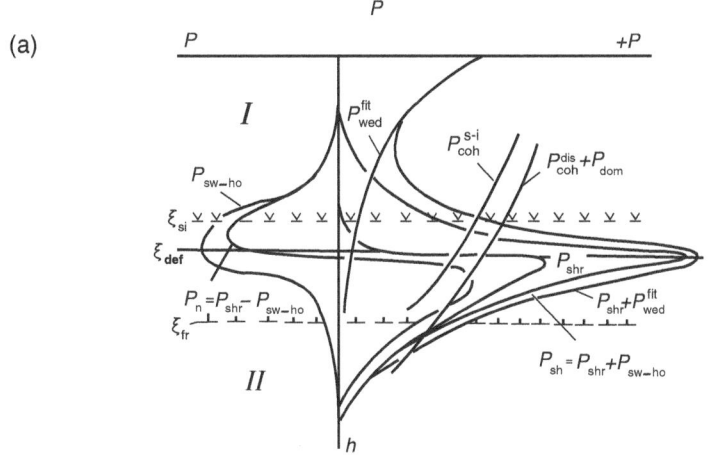

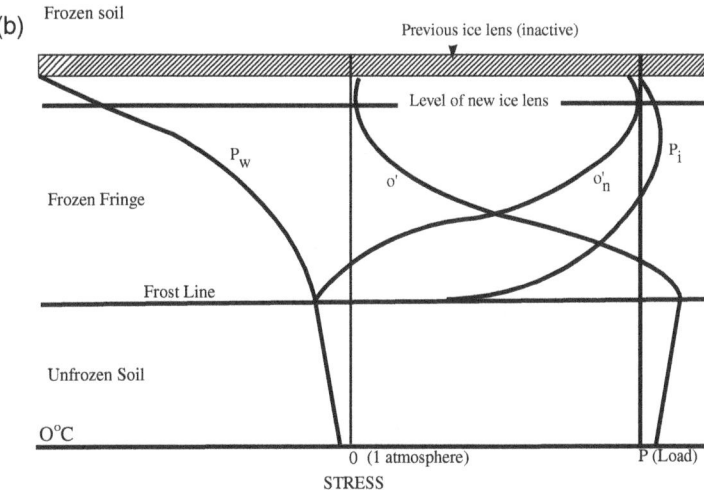

Figure 4.3 The complex stress states in freezing soils: two views. Horizontal axis: stress; Vertical axis depth (a) (above) From Ershov, 1998. Stresses of shrinkage P_{shr} swelling P_{sw}, heave P_{ho}, horizontal shear P_{sh} and normal volumetric-gradient stresses P_n and the wedging pressure of film water P^{fil}_{wed} local cohesion P_{coh} and cohesion at the boundary "soil-ice layer" P^{g-i}_{coh}; ξ_{fr} ξ_{sl} ξ_{def} are boundaries of freezing, visible segregated ice formation and change in deformation direction in the sample, respectively; I, II are frozen and thawed parts of the ground, respectively. (b) (below) From Williams and Smith, 1991, after Miller 1978. $P_i P_w$ - pressures of ice and water; p - load; σ'- effective normal stress (reaction stress of soil matrix) σ_n - neutral stress (pore contents).

sharp change in the soil environment, the microstructure will again tend to stabilize. The fate of the oil will be largely controlled by the soil with this modified structure and fabric.

Ground Surface Changes: Monitoring of Changes by Airborne and Spaceborne Remote Sensing

It is clear that detailed monitoring of spill sites is most effectively performed through *in situ* measurements. However, where potential spill sites are distributed over very large areas with difficult access, airborne or spaceborne remote sensing techniques have a potentially valuable role in identifying and monitoring changes in surface conditions. The main approaches are (1) direct detection of the presence of a free oil surface; (2) detection of changes in thermal regime; (3) monitoring of changes in vegetation characteristics; and (4) monitoring topographic changes. Spillage of oil is immediately obvious on inspection of the ground surface and becomes more so as vegetation dies. Oil penetrates the ground and runoff may also occur, while free oil may persist in hollows or depressions. Although direct detection of an oil surface would be the most satisfactory of these approaches, it is militated against by the potentially short time period for which it is possible, and indirect measurements are therefore also likely to play an important role.

The longer-term effects arising from the disturbance of the ground-surface energy balance, modifications of soil microstructure, and thus of thermal and thermodynamic properties of the soil will cause progressive changes in ground surface relief, drainage, subsequent vegetation, etc. Over long periods, development of karst topography quite often occurs, with extensive subsidences of ground, slope failures, flooding, and modification of drainage patterns. Long before this stage is reached, however, related effects on a much smaller scale are to be expected. The nature, distribution, and rate of development of surface changes reflect the interaction of the oil and the ground, that is, the site-specific modifications of ground properties and behavior discussed in the previous sections.

The manner in which these changes arise from the introduction of the oil needs to be fully understood. Sufficiently detailed monitoring of the ground surface can then be expected to reveal sequential patterns of change capable of interpretation in this way and that will allow prediction of the future course of events. Such prediction opens the way to devising the most appropriate remedial response.

DETAILED RESEARCH REQUIREMENTS: EXPERIMENTAL DETERMINATIONS AND GROUND MONITORING PROCEDURES

Procedures for determination of thermal, thermodynamic, and hydraulic properties and other quantities for freezing soils are now quite well established. These procedures take into account the special nature of freezing soils and should be fairly easily adaptable for samples polluted by oil.

Laboratory Tests: Bench Scale

Many properties can be investigated in bench-scale experiments. Very precise control of temperature is important because of the extremely temperature-dependent nature of the properties, especially at temperatures just below the freezing point.

The most fundamental requirement is for information on the proportions of ice and water (cf. Figure 4.1) following contamination. Small samples can be investigated. Similarly, determinations of heat capacity and thermal conductivity or investigation of the effects of freezing and thawing on soil microstructure can be carried out using fairly small samples. Indeed, determinations of the proportions of ice and water as a function of freezing temperature may be sufficient for calculation of the heat capacity and conductivity over a range of temperatures.

Experiments at Near Natural Scale

Studies of moisture migration and continuing heave or modeling the release of oil at the ground surface can be performed with advantage in larger facilities. It is particularly difficult to maintain temperature gradients in small samples as slight as those commonly measured in the ground (for example 1°C m^{-1} and less). A solution is to operate with quite large soil masses, a cubic meter or more, where control of surface temperatures leads to the gradual establishment of the desired temperature gradient, which is then quite stable.

The Geotechnical Science Laboratories, working jointly with French researchers, have for many years operated a controlled environment facility with a soil bed of 18 × 8 × 2 m. Used first for extensive studies of freezing around pipelines (Geotechnical Science Laboratories, 1993, 1995), the facility has also allowed investigation of slow migrations of moisture (continuing frost heave), of soil resistance, of movements of pollutants (Geotechnical Science Laboratories, 1994 and Winnicky, Chapter 10, this volume) and of soil microstructures. Controlled environment facilities on this scale allow experiments that can model reality closely. Yet because it is possible to control soil lithology and other characteristics, temperature, and groundwater conditions, it is possible to devise experiments that enable the effects of particular factors to be isolated.

A further possibility is the carrying out of experiments under field conditions. Unlike a controlled-environment facility, however, the factors of weather, soil moisture and thermal conditions, and lithology all vary so greatly that it is often difficult to devise experiments that provide clear information on the process or effect being examined.

Whichever procedures are used for experimentation, mathematical modeling on the basis of the established quantities will be a further step in the site-specific analysis of the oil-ground interaction (Corapcioglu and Panday 1993). Modeling of the various thermal, thermodynamic, hydrodynamic, and mechanical aspects of frozen-ground behaviour for specific situations has been substantially developed recently (Fremond and Mikkola, 1993; Ladanyi and Shen, 1989, 1993; Razaqpur and Wang, 1995).

Remote Sensing of Field Conditions

For large spills covering several square kilometers, such as the Komi oil spill in Russia (e.g., Vilchek and Tishkov 1997), satellite-based, remote-sensing observations have potential for the detection and monitoring of the effects especially in permafrost regions, though this potential still requires some experimental verification in field studies. For small spills covering only some tens of square meters, remote sensing has a different meaning: ultimately one may be concerned with visual observations of the ground surface. Such observations at regular intervals, coupled with a good background knowledge of frozen-ground behavior, may also enable prognoses of the future movements and fate of the spilled oil.

The best-known satellite data are those providing imagery in the visible and near-infrared regions of the electromagnetic spectrum, for example, by the Landsat and SPOT satellite series. Such imagery approximates the response of photographic color and false-color infrared photography, with a reduced spatial resolution of typically 10 to 30 m, but a correspondingly increased spatial coverage (each image covers an area typically 100 to 200 km square). The functional distinction between airborne and spaceborne imagery is becoming blurred with the advent of spaceborne systems offering spatial resolutions of the order of 1 m. Spaceborne and airborne nonphotographic data are digital and calibrated and can be used for quantitative analysis. Vegetated surfaces yield characteristic signals due to (1) absorption of visible wavelengths by plant pigments and (2) scattering of near-infrared wavelengths by cell structures. Many algorithms, for the classification of vegetation and the estimation of biomass, have been derived for lower latitudes and are now being modified for use at high latitudes (e.g., Rees and Williams, 1997). Such algorithms may be expected to reveal the extent of oil pollution through its toxicity to plant species.

Exposed oil surfaces are also expected to be readily discriminated from the "background" vegetation, or from snow or bare ground, through the characteristically low reflectance values of the free-oil surface. (It should be noted, though, that open water surfaces can have a similar low reflectance, so that it is necessary to have prior knowledge of the positions and extents of water bodies.) Similarly, oil-soaked ground may be detectable. It should be possible to infer soil type from such visible/near-infrared observations over unvegetated ground.

The increased availability of "hyperspectral" remote-sensing systems, offering hundreds or thousands of spectral channels across the visible and near-infrared region of the electromagnetic spectrum, substantially increases the scope for discriminating vegetation species and state of health, and also for identifying exposed oil. Part of the reason for this increased potential is the possibility of detecting fluorescence phenomena in both vegetation (chlorophyll) and oil. Hyperspectral systems have already been flown on airborne missions and are planned for spaceborne deployment in the near future.

Other remote-sensing techniques are also available. Observations in the thermal infrared (≈ 10 μm) part of the electromagnetic spectrum yield, with suitable field validation, surface temperatures. If the observations are made at appropriate times of day, the diurnal surface temperature fluctuation, and hence the thermal inertia, can be determined. This may be related in a reasonably straightforward way to the concentration of oil in the ground.

Indeed the relationship of the ground thermal regime to the surface-energy exchange, which is itself substantially dependent on the nature of the surface and vegetative cover, strongly supports this. The thermal regime of the ground is, in turn, a powerful indicator of the condition of the ground. Both the visible/near-infrared and the thermal infrared techniques described above are dependent on cloud-free conditions and daylight. Neither of these can be guaranteed at the times at which oil spills may occur in permafrost environments. A third possibility is provided by synthetic aperture radar (SAR) systems, which can achieve spatial resolutions of the order of 10 m and which are independent of daylight or cloud cover conditions. The variable that is measured by such systems is the radar backscattering coefficient, which is determined by the dielectric and geometric properties of the target material in a fairly complicated way that inhibits interpretability. Streck and Wegmüller (1997) have demonstrated that spaceborne SAR is capable of delineating pipeline structures themselves, as well as complementing other methods of deriving land cover. Although research is now being carried out at various centers, including the Scott Polar Reseasrch Institute, on the appearance of oil spills in permafrost environments, little has been published. However, Allen and Wilson (1995) have shown that the Komi oil spill is detectable in SAR imagery of the Usinsk area, and we believe that this represents a very promising approach to the detection and definition of the extent of such spills, at least in the earlier stages when the oil spill presents a surface above the ground (Figure 4.4). As with the use of optical/near-infrared data, however, oil surfaces can be confused with water surfaces, so again it is necessary to have prior knowledge of the distribution of water bodies.

Comparisons of the Usinsk spill as delineated by satellite SAR imagery (Figure 4.4, November 2) with ground-survey data (November 29) suggest (unpublished research carried out at the Scott Polar Research Institute) that the spilled oil is visible only when its depth exceeds approximately 30 cm. This effect is currently under investigation. Even if it is true that the SAR method does not respond to oil depths less than about 30 cm, reasonably accurate spill volumes could still be obtained from the analysis of satellite imagery if a sufficiently detailed Digital Evaluation Model (DEM) of the spill site were available. Such a DEM would need to have a vertical resolution of perhaps 10 cm, considerably more detailed than is generally available, and would therefore probably need to be generated specifically for a potential spill site (for example, as a strip along the course of a pipeline).

It is possible that SAR imagery will also be able to reveal the presence of oil-soaked ground without a free-soil surface, through a change in the dielectric properties of the ground (Topp et al., 1980; Patterson and Smith, 1985), and vegetation damage through changes in the structure of the vegetation canopy. These possibilities will be investigated. It is, however, already well known that oil slicks on water surfaces can be detected through the damping effect they have on surface waves and ripples (e.g., Martinez and Moreno, 1996). Thus, the introduction of spilled oil into larger bodies of liquid water should be readily detectable using SAR imagery.

Satellite remote sensing approaches should also be able to detect and quantify the gross long-term consequences of ground disturbance. The formation of karst

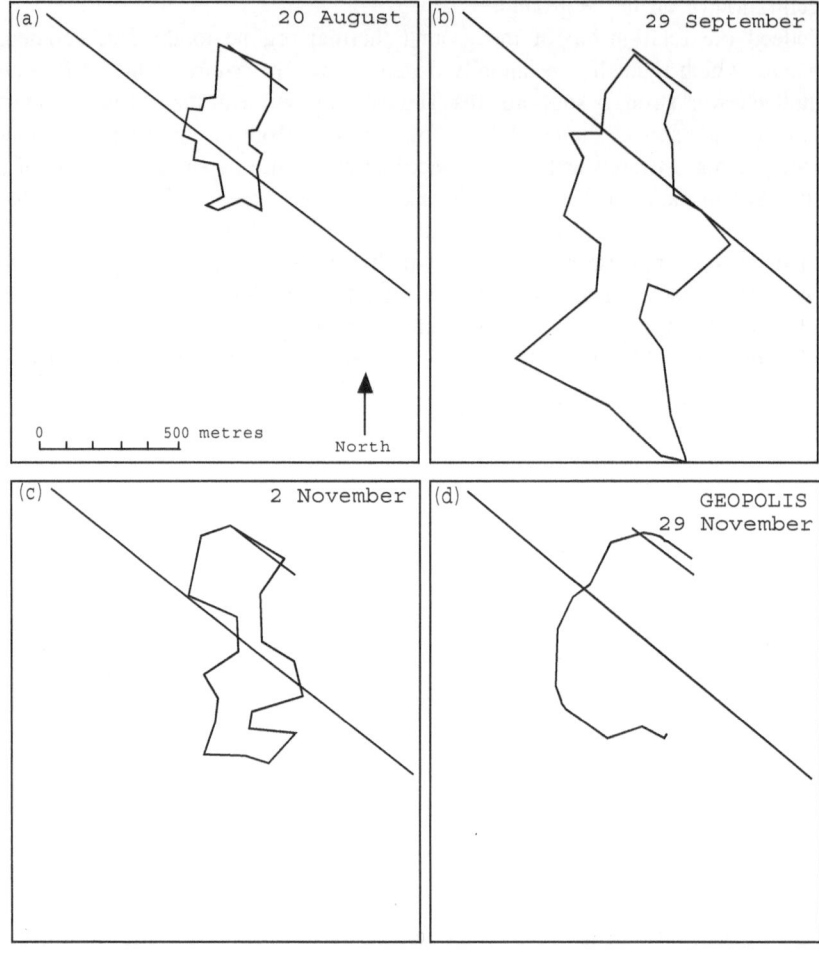

Figure 4.4 Delineation of the Usinsk oil spill, Komi Republic Russia, 1994, from analysis of ERS-1 synthetic aperture radar (SAR) images. (a), (b), and (c) show the outline of the detectable parts of the spill on 20 August, 29 September, and 2 November respectively; (d) shows the boundary of the spill determined by *in situ* measurement on 29 November. The heavy lines show the positions of the pipelines, which can also be determined from the SAR imagery.

topography should be readily detectable in high-resolution, visible-wavelength imagery, especially at low solar elevations when topographic detail is enhanced in such imagery. Flooding can be detected in a number of different types of imagery through different mechanisms. Drainage patterns can be monitored through the characteristic signal of water in near-infrared imagery and perhaps inferred from the topographic detail present in visible-wavelength imagery. These techniques have, however, received little attention in cold regions, and extensive investigation and validation is still required.

In summary, we can state that airborne and spaceborne remote-sensing methods have substantial potential to characterize oil spills and their effects, albeit principally through indirect measurements. This potential is barely tested, and the purpose of the preceding section has been to outline what seem to us to be the most promising lines of attack. It is very likely that this approach will be most effective when pursued in the environment of a geographic information system (GIS), providing a suitable means of data integration. Coarser-scale data from satellite remote sensing can also be incorporated into such a GIS. The optimal design of suitable GIS structures requires investigation.

COORDINATION OF FINDINGS

Essential to the research strategy proposed is the combining of knowledge of the interaction of oil and freezing ground with monitoring of ground surface changes and their interpretation. No group of researchers at any single institution has the expertise to carry out the entire range of investigations required. Certain of these investigations can be carried on largely independently and indeed may provide information urgently required on specific topics (for example, the decline of permeability to oil of frozen ground with falling negative temperature). But the thrust of this proposal is that by combining the expertise of several institutions the work will proceed in a comprehensive fashion to give a detailed picture of the effect of the spilled oil on the total system, atmosphere/ground surface layer. The stage will ultimately be reached at which informed assessments can be made of oil spills, including prediction of the longer-term developments. This will mean that rational and considered cleanup procedures can then be applied with maximum cost-effectiveness.

PARTICIPATING GROUPS

The strategy proposed here is based upon the experience of the Geotechnical Science Laboratories of Carleton University in field and laboratory studies of frozen ground and of the Scott Polar Research Institute, Cambridge University, U.K., in remote sensing in the Arctic. Both groups have benefited from links with Russian researchers in the Departments of Geocryology and Geography of Moscow State University who, it is intended, will be involved in future work in this connection. It is planned that, with the extension of facilities at Carleton University, researchers from Norway and Finland will also participate in experimental studies and assist in analysis. Both the lead institutions have long-standing connections with these countries, while the Carleton workers continue to have links with their French colleagues from the earlier joint projects in France.

Cleanup costs for cold-regions ground contamination are so large that there is need for extensive research and technological development. Our own limited experience, we believe, indicates the appropriate strategy.

FUNDING AND EQUIPMENT

Several European sources are currently funding work that pertains to the strategy outlined. Limited Canadian federal government funding has facilitated the work on microstructures and on experimental pollution studies. Additional cold-room space has recently been installed at Carleton University for the testing of large soil specimens and, under arrangements with Agriculture Canada, equipment for investigation of soil microstructure has also been located here. The rate at which investigations proceed is likely to depend largely on the perceptions of the environmental industry of the need for and the financial advantages of improved remedial techniques for oil spills.

CONCLUSION

Gaining knowledge through experimentation and modeling of the effects of contaminants on the complex thermodynamic (phase change) relations of freezing soils and in particular of their effects at the microscopic level will open the way to prediction by remote sensing of the course of changes due to ground contamination. Such predictions will provide the basis for selection of the most rational and cost effective remedial procedures on a site-specific basis.

REFERENCES

Allen, P. and S. Wilson. NRSC Uses ERS-1 SAR Data to Monitor Komi Oil Spill. *Earth Observation Magazine*, 41–43, 1995.

Andersland, O.B. and D.M. Anderson. *Geotechnical Engineering for Cold Regions*. McGraw-Hill, New York, 1978, 566 pp.

Boitnott, G.E., I.K. Iskandar, and S.A. Grant. The Use of Frozen Ground Barriers for Containment and *In Situ* Remediation of Heavy-Metal Contaminated Soil, in *Intl. Symposium, Physics, Chemistry and Ecology of Seasonally Frozen Soils*. CRREL, Fairbanks, Alaska. Spec. Rep. 97–10, 409–416, 1997.

Chamberlain, E.J. and A.J. Gow. Effect of Freezing and Thawing on the Permeability and Structure of Soils. *Eng. Geol.* 13, 73–92, 1979.

Corapcioglu, M.Y. and S.M. Panday. Simulation of Hydrocarbon Spills in Permafrost. *Proceedings, Sixth Intern. Conf. Permafrost*. Beijing, 100–103, 1993.

Dash, J.G., H.Y. Fu, and R. Leger. Frozen Soil Barriers for Hazardous Waste Confinement. *Ground Freezing 97, Proc. Int. Symp. Ground Freezing and Frost Action in Soils*, Balkema, Rotterdam, 375–380, 1997.

Ershov, E.D. General Geocryology (English translation from the Russian, Nedra 1990, 559 pp.). Cambridge University Press, New York, 1998, 580 pp.

Fremond, M. and M. Mikkola. Thermomechanical Modelling of Freezing Soil, in *Gas Pipelines, Oil Pipelines and Civil Engineering in Arctic Climates*. Proceedings of seminar held in Caen and Paris. Geotechnical Science Laboratories, Carleton University, Ottawa, 48–60, 1993.

Geotechnical Science Laboratories. Various articles in *Gas Pipelines, Oil Pipelines and Civil Engineering in Arctic Climates*. Proceedings of Seminar held in Caen and Paris. Geotechnical Science Laboratories, Carleton University, Ottawa, 1993, 173 pp.

Geotechnical Science Laboratories. Study of Movement of Hydrocarbons through Freezing and Thawing Soils (Dew Line Cleanup): Final Report to Department of National Defence, Canada. 2 volumes, 1994, 80 pp. and 244 pp.

Geotechnical Science Laboratories. Publications, Reports and Theses, Supplement: Pipeline Ground Freezing Proj. (listing of references only), 1995, 21 pp.

Greschischev, S., A.V. Pavlov, and V.V. Ponomarev. Changes in Microstructure of Fine-Grained Soils Due to Freezing. *Permafrost and Periglacial Processes*, 3(1): 1–10, 1992.

Iskandar, I.K. and F.H. Sayles. Ground Freezing for Containment of Hazardous Wastes. Engineering Aspects, in *Intl. Symposium, Physics, Chemistry and Ecology of Seasonally Frozen Soils*. CRREL, Fairbanks, Alaska. Spec. Rep. 97–10, 361–369, 1997.

Johnston, G.H. *Permafrost Engineering, Design and Construction*. Wiley, Canada. 1981.

Ladanyi, B. and M. Shen. Mechanics of Freezing and Thawing in Soils. *Proc. Frost in Geotechnical Engineering*. 1 (Finland) 73–103, 1989.

Ladanyi, B. and M. Shen. Freezing Pressure Development in a Buried Chilled Pipeline, in *Proceedings, 2nd International Symposium on Frost in Geotechnical Engineering*. Rotterdam, Balkema. 23–33, 1993.

Martinez, A. and V. Moreno. An Oil Spill Monitoring System Based on SAR Images. *Spill Science and Technology Bulletin*, 3, 65–72, 1996.

Miller, R.D. Frost Heaving in Non-Colloidal Soils, in *Proceedings Third International Conference on Permafrost*. Edmonton. 708–713, 1978.

Patterson, D. and M.W. Smith. Unfrozen Water Content in Saline Soils Due to Freezing. *Canadian Geotechnical Journal*, 22(1), 95–101, 1985.

Razaqpur, A. G. and D. Wang. A Simplified One-Dimensional Frost Heave Model, in *Proc. Engineering Mechanics Symposium*. Canadian Society for Civil Engineering. 102–108, 1994.

Rees, W.G. and A.P. Kapitsa. Industrial Pollution in the Kol'skiy Poluostrov, Russia. *Polar Record*, 30(174), 181–188, 1994.

Rees, W.G. and M. Williams. Monitoring the Impact of Atmospheric Pollution on Tundra Vegetation in the Russian Arctic: Interim Report, SPRI Technical Reports in Remote Sensing, No. 5, 1995.

Rees, W.G. and M. Williams. Monitoring Changes in Land Cover Induced by Atmospheric Pollution in the Kola Peninsula, Russia, using Landsat-MSS Data. *International Journal of Remote Sensing*, 18, 1703–1723, 1997.

Solntseva, N.P. and O.A. Guseva. Distribution of Oil and Oil Products in Soils of Tundra Landscapes within the European Territory of Russia (ETR) Soil, in *Intl. Symposium, Physics, Chemistry and Ecology of Seasonally Frozen Soils*. CRREL, Fairbanks, Alaska. Spec. Rep. 97-10, 449–453, 1997.

Streck, C. and U. Wegmüller. Investigation of ERS SAR Data of the Tandem Mission for Planning and Monitoring of Siberian Pipeline Tracks, in *Proceedings of the Third ERS Symposium on Space at the Service of Our Environment*, Florence, March 17–21, 1997. ESA SP–413, 441–447, 1997.

Topp, G.C., J.L. Davis, and A.P. Annan. Electromagnetic Determination of Soil Water Content: Measurements in Coaxial Transmission Lines. *Water Resources Research*, 16(3), 574–582, 1980.

Vilchek, G.E. and A.A. Tishkov. Usinsk Oil Spill: Environmental Catastrophe or Routine Event? in *Disturbance and Recovery in Arctic Lands*. R.M.M. Crawford, Ed., Kluwer, Dordrecht. 411–420, 1997.

Virtual Conference on Contaminants in Freezing Ground (contains comprehensive bibliography) See: http://www.freezingground.org/vc. Scott Polar Research Institute, Cambridge, U.K., 1998.

Washburn, A.L. *Geocryology: A Survey of Periglacial Processes and Environments.* Arnold, London. 1979, 406 pp.

White, T. L. Cryogenic Alteration of Clay and Silt Microstructure. Implications for Geotechnical Properties. Ph.D. thesis presented to Carleton University, Ottawa, Canada, in partial fulfillment of the requirements for the degree of Doctor of Philosophy, 1995.

White, T.L. and P.J. Williams. *Cryogenic Alteration of Frost-Susceptible Soils.* in *Proc. 7th Intern. Symp. Ground Freezing*, Nancy, France, 1994, 17–24.

Williams, P.J. Research Strategies for Development of Predictive and Remedial Measures for Oil Spills in Permafrost Regions, in *Proc. Wkshp. Technologies and Techniques for Hydrocarbon Remediation in Cold and Arctic Climates*. Royal Military College of Canada, Kingston, 1996.

Williams, P.J. and M.W. Smith. *The Frozen Earth. Fundamentals of Geocryology.* Cambridge University Press, New York, 1991, 304 pp.

Williams, P.J., W.G. Rees, and T.L. White. A New Approach for Ameliorative Strategies for Hydrocarbon Spills. Report to European Research Office, U.S. Army. Geotech. Sci. Labs. Int. Rep. No. 69. 1996, 24 pp.

SECTION II

Example Applications of Models for Cold Regions

Basic Guidelines for Conducting Groundwater Modeling Efforts to Meet Environmental Requirements

G.R. Bruck

CONTENTS

INTRODUCTION

Experience has shown that there are two recurring problems with groundwater modeling efforts when they are used for compliance with environmental goals. Many times, decision makers and managers are not satisfied with modeling products they have requested and, similarly, modelers are often dissatisfied with the guidance they receive for conducting modeling work. This presentation attempts to offer some helpful information to both groups.

Perhaps the single most effective tool in avoiding these problems is to draft a simple one-paragraph statement that outlines the modeling objectives along with some basic consideration of how the results will be used. Using this simple statement as a basis, modelers and decision makers can make a realistic appraisal as to whether a system can be modeled correctly and how much effort the model will require. Such communication is especially critical *before* the initial stages of the work begin. Although this may seem like an elementary step, in too many instances it does not occur.

From one perspective, it is perhaps appropriate to think of purchasing a "modeling effort" as a process that is not unlike buying a car. It is essential to get a clear picture, early on, of what the model can and cannot do. In this regard, an assessment of the model's accuracy is very important. It is also critical to determine exactly how the results will be presented, and who is the model's intended "audience." Carefully think through how the results might be interpreted by the end-users, since, unfortunately, there are many ways that good results can be misinterpreted by the general public or end-users looking for a specific outcome.

There are two excellent publications that can assist project managers and decision makers in writing a statement of modeling objectives and assessing their modeling needs. The first, an EPA document, *Assessment Framework for Groundwater Model Applications* (EPA 1994a), contains a simplified flowchart of the typical groundwater modeling process, along with an excellent series of questions that will help all parties decide what kind of modeling effort is appropriate. The sequence of questions can be used as a checklist. The second publication, *Groundwater Modeling Compendium, 2nd Edition* (EPA 1994b), incorporates a wealth of information on many of the commonly used groundwater models, including information on estimating and managing modeling costs. It also contains examples of real-world problems where models were used, including assessments of how successful the efforts were in meeting their objectives. It is a good source of example problems, case studies, model descriptions, and fact sheets on some of the more commonly used models.

In an attempt to remedy the second issue mentioned above, the lack of "ground rules" for modelers, the following guidelines are offered. These guidelines should provide some insights as to how the technical staff in the EPA Region 10 office review and evaluate modeling efforts. Several staff members in our office developed them, and they reflect a wide range of modeling experiences. Most of the ideas here were compiled from the technical reviews of modeling efforts with which we have been involved.

GUIDELINES ON HYDROGEOLOGIC MODELING

The intent of these guidelines is to provide a basic list of elements to consider when using groundwater and surface water models. The information presented here is based on experiences in the EPA Region 10 office (Seattle) with environmental projects that involve Superfund, RCRA, NPDES, NEPA, TSCA, SDWA, and FIFRA programs.

For our purposes, the term *model* includes any predictive exercise that attempts to describe the future movement of groundwater, surface water, or contaminants. By applying these criteria, efforts such as qualitative or graphical depictions, "back-of-the-envelope" calculations, hand calculator exercises, and computer simulations should also be viewed as models. However, the principal targets for these guidelines are complex mathematical models or computer calculations.

The variability of earth materials, groundwater and surface-water systems, contaminant sources, types, and project goals *requires* that each site be evaluated on a site-specific basis. The type and complexity of the model that will be used *must* reflect those variations. The references noted below, though not exhaustive, contain examples of mathematical approaches and computer codes that have been used in a variety of situations.

The approach taken in these guidelines attempts to ensure that modeling tasks involve a clear progression of thought, starting with the question posed, proceeding to a conceptual model of the system, and followed by computer runs of the most appropriate scenarios, which finally result in an answer to the original question. The end products should be based on the best possible understanding and site data and *must include* both an uncertainty analysis and an acknowledgment of the model limitations and data gaps. It should also be recognized that the iterative nature of modeling will commonly indicate the need for additional site data or modeling scenarios.

Two points that are emphasized in these guidelines are the information that should be supplied as the protocol in a work plan or scoping document and, second, the preferred use of computer codes that are in the public domain.

Modeling Protocol

A modeling protocol should be developed and provided in a work plan or scoping document. This protocol should be very specific in identifying the goals of the model. It should also provide planning information about how much effort will be needed to complete the model. In the modeling protocol, all data for model input parameters and boundary conditions that need to be collected during the field work should be identified and their method of collection described in the sample plan. The following items should be included in the modeling protocol:

1. The goals of the modeling should be stated in simple terms that will allow easy comparison to modeling results so that a determination can be made as to whether goals were achieved.
2. A site-specific conceptual model of the hydrogeologic system should be presented in graphical form (including scaled maps and cross sections) with associated description and discussion. This must be based on available data. The conceptual model should include the mathematical relationships that are used to describe the principal hydrogeologic processes of concern. Any data gaps, assumptions, and uncertainties should be described in the conceptual model.
3. The technical requirements needed to achieve the goals should be discussed. Major factors include the following:
   - Numerical vs. analytical approach
   - Two- or three-dimensional requirements
   - Simulation of the unsaturated zone
   - Multiphase flow
   - Tidal influence
   - Requirements for simulation of dispersion, retardation, and degradation
   - Decay and chemical reactions

4. If computer modeling is needed, the appropriate computer codes should be identified. The ability and limitations of the proposed computer codes to meet the conceptual model requirements should be discussed. It is critical to match the proposed code to the site conditions and the availability of data, including good justification for the input parameters.
5. Boundary and initial conditions and other input parameters should be identified, and the following procedures for the model calibration and uncertainty analysis should be described:
   • The input parameters that will have assumed values, and the basis for those assumptions
   • The input parameters that will have values derived from site-specific information
   • The parameter estimation techniques and associated uncertainties
   • The plan for parameter sensitivity analysis to compare model sensitivity with variations in input parameters

A brief description of data collection methods (or a reference to a section in a sampling and analysis plan) should be provided. The quality of and variability associated with input parameters derived from site-specific data also should be described.

6. Model documentation should incorporate the Elements of Technical Analysis listed in Table 5.1. Two key elements of a model report should be noted:
   • Uncertainty analysis
   • Copies of critical model inputs
   The uncertainty analysis can be either qualitative or quantitative, depending on the scope of the modeling. A computer modeling report also should include an appendix with copies of critical model input and output files. The report should be accompanied by a floppy disk containing all input files used in the evaluation.

Computer Codes

The EPA Region 10 hydrogeologic staff advises the use of public domain computer codes over proprietary codes for the following reasons:

1. The Region 10 technical staff and EPA consultants are familiar with a number of public domain computer codes. In cases where the code has not yet been used in Region 10, there is a good chance that other EPA regional or national modeling experts will be familiar with it, if it is a public domain model.
2. In an oversight capacity, the Region 10 technical staff or EPA consultants generally do not have time to become familiar with the operation of proprietary computer models. This is an especially significant factor at sites with relatively short time schedules established for EPA review.
3. The Region 10 staff and EPA consultants do not have free access to proprietary codes and must purchase them or make special arrangements for their use in a proprietary manner. This is unjustifiable considering the many available high-quality public domain codes that have undergone extensive peer review and validation, and that have been accepted in legal arenas.
4. Modeling projects under most EPA programs are subject to public review. Proprietary computer models are not readily available to the public. Those people in the public sector who are familiar with modeling or who have access to hydrogeologic

Table 4.1 EPA Region-10 Office of Environmental Assessment Elements of Technical Analysis for Modeling

I. Management Objectives
  Scope of the problem
  Technical analysis objectives as they relate to management objectives
  Level of analysis required
  Level of confidence required
II. Conceptual Model
  System boundaries
  Important time and length scales
  Important processes
  System characteristics
  Source characteristics
  Available data sources (quality and quantity)
  Data gaps
  Data collection programs (quality and quantity)
III. Choice of Technical Approach
  Rationale for approach in context of management objectives and conceptual model
  Reliability and acceptability of methodology
  Important assumptions
IV. Parameter Estimation
  Data used for parameter estimation
  Rationale for estimates in the absence of data
  Reliability of parameter estimates
V. Uncertainty/Error
  Error/uncertainty in input and boundary conditions
  Error/uncertainty in loadings
  Error/uncertainty in specification of environment
  Structural errors in methodology (e.g., effects of aggregation or simplification)
VI. Results
  Tables of all parameter values used for analysis
  Tables or graphs of all results used in support of management objectives or conclusions
  Accuracy of results
  Conclusions of analysis in relationship to management objectives
  Recommendations for additional analysis, if necessary

consultants should have the opportunity to evaluate the basis, validity, and sensitivity of modeling results.

5. By discouraging or limiting technical review, the use of proprietary computer codes tends to promote predictive calculations as a black-box science and serves to limit public evaluations of the hazards to health and the environment.

REFERENCES

Anderson, M. and W. Woessner. *Applied Groundwater Modeling: Simulation of Flow and Advective Transportation.* Academic Press, New York, 1992.

Bond, F. and S. Hwang. *Selection Criteria for Mathematical Models Used in Exposure Assessments: Groundwater Models,* U.S. Environmental Protection Agency. Pub. No. EPA/600/8-88/075, 1988.

Javandel, I., C. Doughty, and C.F. Tsang. *Groundwater Transport: Handbook of Mathematical Models,* Water Resources Monograph No. 10, American Geophysical Union, Washington, D.C., 1984.

National Research Council, *Groundwater Models: Scientific and Regulatory Applications,* National Academy Press, Washington, D.C., 1990.

U.S. Environmental Protection Agency, *Assessment Framework for Groundwater Model Applications,* Solid Waste and Emergency Response (5103) Directive No. 9029.00, EPA Pub. 500-B-94-003, 1994a.

U.S. Environmental Protection Agency, *Groundwater Modeling Compendium,* 2nd ed., Model Fact Sheets, Descriptions, Applications and Cost Guidelines. Solid Waste and Emergency Response (5103), Pub. No. EPA 500-B-94-004, 1994b.

U.S. Environmental Protection Agency, *Agency Guidance for Conducting External Peer Review of Environmental Regulatory Modeling,* Agency Task Force on Environmental Regulatory Modeling (ATFERM), Office of the Administrator (1102) Pub. No. EPA-100-B-94-001, 1994c.

Van der Heijde, P., Y. Bachmat, J. Bredehoeft, B. Andrews, D. Holtz, and S. Sebastian. *Groundwater Management: The Use of Numerical Models,* Water Resources Monograph No. 5, American Geophysical Union, Washington, D.C., 1985, 180 p.

Van der Heijde, P., A. El-Kadi, and S.A. Williams. *Groundwater Modeling: An Overview and Status Report,* U.S. Environmental Protection Agency, Washington, D.C., EPA/600/2-89/028, 1988.

CHAPTER 6

Hydrogeoecological Problems in Developing the Diamond-Bearing Deposits in Northern Regions of Russia

V.A. Mironenko and F. G. Atroshchenko

CONTENTS

INTRODUCTION

As in many countries, mining operations in Russia have been widely viewed as a source of environmental disruption and degradation. Future measures to prevent or remediate these effects are expected to cost billions of dollars. Apart from inadequately sophisticated methods for studying and predicting environmental processes, the main cause of the present situation is the lack of integrated environmental monitoring and control procedures at mining operations. This is particularly true of cold regions sites, which some view as more vulnerable to environmental disturbance. An analysis of prospecting and development of diamond-bearing deposits of Yakutia, Siberia, and the Arkhangelsk region in the European portion of Russia, where we have conducted comprehensive hydrogeological studies for a number of years, should be of interest for environmental protection in mining operations generally and particularly for those in cold regions.

1-56670-476-6/00/$0.00+$.50
© 2000 by CRC Press LLC

DEPOSITS OF THE YAKUTSK DIAMOND-BEARING PROVINCE

The "Mir" pipe deposits (Yakutia) are being exploited under complex mining and hydrogeological conditions. Brines (with concentrations as high as 120 g/L) below the permafrost (Figure 6.1) result in inflows on the order of 1000 m³/hour into the mine pit. Prior to 1988, inflow was prevented primarily by lowering the level of the surrounding groundwater. The resulting saline waters accumulated during the winter in temporary containers and basins, and they were then discharged in summer to the river network. The amounts of saline water discharged at any one time were adjusted so that the maximum permissible salt concentrations would not be exceeded in the control stretch of the Vilyuy River.

As the mine pit was deepened, the environmental situation was aggravated as brine flows into the mine pit increased. Accordingly, an alternative technology was developed to reduce sharply the inflow of water into the mine pit and to eliminate the releases of drainage waters into the adjoining river network. The principal elements of the technology were an insulating curtain (IC) and a system for back-injecting drainage water into the brine horizon. After the stability of the open-pit slopes was estimated and the hydrostatic and hydrodynamic forces were considered, the following version of the insulating curtain was adopted for implementation.

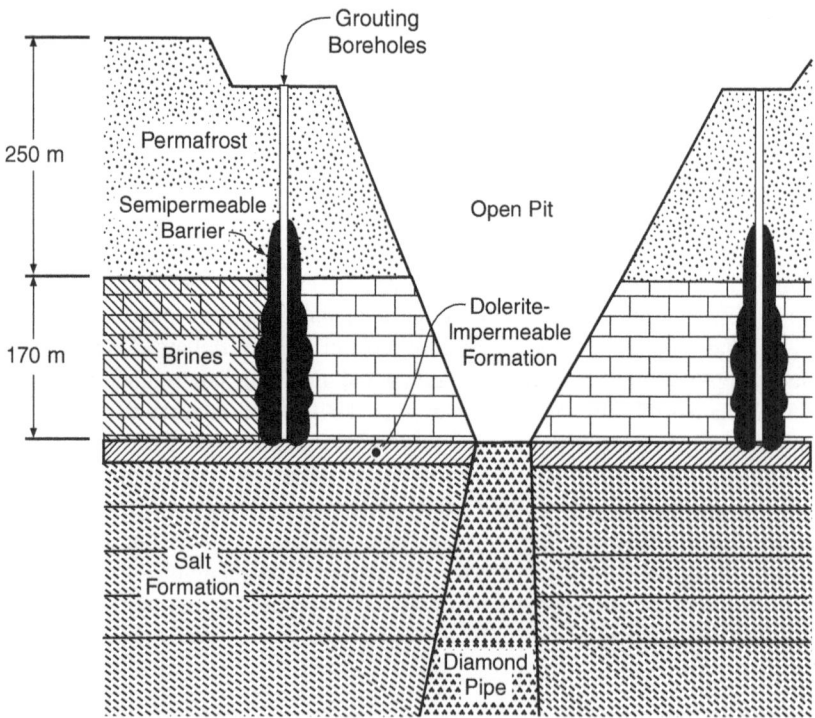

Figure 6.1 Schematic profile at the "Mir" mine.

Boreholes were to be drilled at 20-m intervals from the top berms of the pit to depths of roughly 350 m, and a clay-cement grouting mixture then will be injected into the main brine-bearing strata. When completed, this grout curtain should have an overall perimeter of around 3100 m, a depth of around 230 m, and a nominal thickness of 32 m. The permeability within the IC should be reduced to 1 to 2 mm/day from approximately 1 m/day. The creation of the IC is expected to reduce the inflow into the mine pit to no more than 250 to 300 m^3/hour without any drainage system. This reduction in inflow would greatly simplify dewatering the mine pit and disposing of the recovered salt water.

Construction of the IC was completed in 1997. Since the drainage system was eliminated, the mine pit was flooded during this construction. After the IC was in place, it allowed the mine to be deepened by approximately 70 to 80 m over 3 to 4 years, after which it was proposed that deeper portion of the diamond-bearing pipe would be extracted by underground mining.

The development of the IC concept for an open-pit mine has been controversial, and several critical questions remain either unanswered or disputed. For example, the presence of even a few relatively permeable "windows" in the IC would sharply reduce its ability to limit inflow. Our modeling exercises have shown that if ungrouted portions occupy only a small percentage of the IC surface, the inflow rates into the mine pit will nearly double — equivalent to increasing the permeabilities of the grouted rocks by a factor of 2 to 3.

The lengthy period of construction and operation of the IC placed strict demands on quality control. The following factors should be noted although their ultimate importance is unclear:

- Three aspects of the site may limit the long-term stability of the IC structure: prolonged contact with the brines; greater hydrodynamic pressure in the IC zone (calculated gradient of flow through the curtain is around 7 to 10); and the likely appearance of permeable "windows" in the IC where concentrated inflow flows may be focused.
- The IC intercepts the frozen rock at its top and the salt-containing strata at its bottom. Thus, the IC may serve to bring brines in contact with the frozen rock, thereby accelerating the melting of the frozen rock.
- There is a possibility of limited shear creep deformations of rock in the IC zone, even when the IC is located beyond the boundaries of possible landslides of the open pit slope.

Accordingly, we have developed a technique for estimating and verifying the quality of the IC construction based on double-tracer experiments in which the injection boreholes and the extraction boreholes are located on opposite sides of the IC.

In addition, we have developed a method for "ice grouting" the IC during construction. An initial set of boreholes is drilled to create the basic IC, and a clay-cement mixture is pumped into them filling the major fractures. A second set of boreholes is then drilled, in which the new boreholes are placed in between boreholes of the first set. Fresh (i.e., nonsaline) water is then injected through the second set of boreholes, filling (and freezing in) the finer pores of the rock.

The proposed method entails preliminary heating of the massif in the ice-grouting zone by injecting warm brine, creating a gradual transitional zone in the massif between brine and freshwater (ice) zones. The heating of the massif, or rather of the external zones of the rock blocks, allows the fresh water to move through practically the entire active fracture space within the zone of influence of excessive hydrostatic pressure in the borehole. In this way, the fresh water not only fills (and then plugs) the fractures free of grouting material within the basic IC but also greatly enlarges its volume. In this case, the enlargement of the IC can be achieved with much lower pressures in the boreholes (often under conditions of free inflow, i.e., independent of the drilling equipment) than when using conventional grouting solutions. This also sharply reduces the rate of residual flow through porous blocks, which are covered on all sides by an impermeable envelope, even along tiny fissures inaccessible for the grouting solution. Finally, a transitional zone eliminates direct contact between the forming ice and the highly concentrated brine solutions, so that the corrosive effect of the latter will be greatly mitigated.

The proposed ice grouting method is appropriate primarily for fractured or fractured-porous rock, in which the rock blocks can be heated at their contact with the fractures while simultaneously maintaining freezing temperatures in the interiors of the rock matrix. This enables fast recooling of the heated fractures and prompt freezing of the warm nonsaline water injected into them.

The ice grouting method is combined with our proposed observation network of boreholes in the body of the IC for long-term monitoring of underground salt water flow and for indentifying zones in which the grouting is incomplete. The shafts of these boreholes are filled with fresh ice; electrical resistance sensors in the borehole can easily detect any melting of ice by contact with leached brines.

Yet another measure of the environment protection is the secondary injection of the pumped-out brines, which is promoted by the presence of regional joints with significantly reduced flow properties. In fact, the largest of these joints is used as a natural semipermeable barrier, separating the back-injection site from the mine pit. The experience in the operation of the injecting boreholes has shown that the back-injection of undersaturated brine into the brine-bearing stratum results in leaching of salt out of the underlying salt stratum near the borehole, and consequently there was an increase in the total water-intake capacity of the injecting boreholes system to 1000 m^3/hour. The backflow of the injected brine into the mine pit (with the IC unfinished at present) amounts to 30 to 35% of the injected recharge.

A different approach to the disposal of subterranean brines is being implemented at the "Udachnaya" pipe deposit. The brine-bearing horizons lie approximately 200 m beneath the permafrost zone (PFZ). These horizons are responsible for an inflow of brines (with a total disolved solids content of over 300 g/L) into the mine pit on the order of 200 m^3/day. The salinity of the brine is too high for it to be pumped into surface-water courses. On the other hand, the low intake capacity of the brine-bearing rock does not allow a secondary injection of the pumped-out brine into them. In this situation, it was decided to return the brine to the PFZ, represented by an interstratification of limestone, marl, argillite,

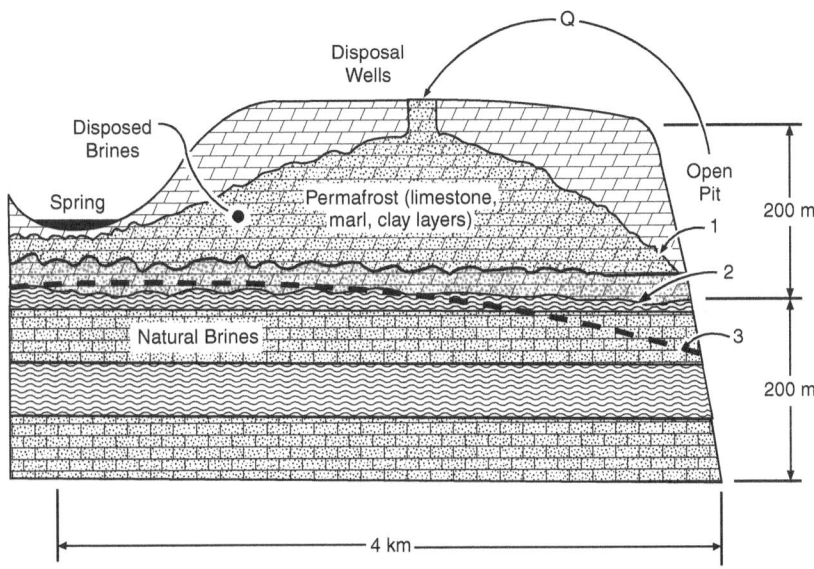

Figure 6.2 Schematic profile at the "Udachnaya" mine. (1) Boundary of spreading of the injected brine; (2) Boundary of permafrost zone; (3) Pressure level in natural brine.

aleurolite, and dolomite, with an average coefficient of permeability of several tenths of a meter per day (Figure 6.2). Leaching the ice, the brines fill in the residual free porosity of the fractured-porous rocks, which is approximately 3 to 4%. The potential free volume capacity of the PFZ portion, which lies below the surface elevations of the water courses, significantly exceeds the overall volume of brine to be pumped out during the entire mining operation. The theoretical possibility of brine disposal into the permafrost zone through surface basins and boreholes situated several kilometers away from the mine pit has been confirmed by a pilot operation, where discharge volumes have been on the order of 100 m³/h for 7 years.

The key question here is the maximal use of the potential capacity of the permafrost zone: If the infiltrating brines quickly reach the first brine-bearing stratum (in the bottom of the PFZ), they will begin to gradually enter the mine pit, since a closed circuit will ultimately be formed between the recharging facilities and the mine pit. In this situation, the effectiveness of the underground disposal is maximized when the brine-recharging boreholes are shallow, allowing the salt solutions to spread out laterally with gradual descending infiltration by leakage through relatively impervious beds that occur sporadically within the permafrost zone. The process is monitored by a system of observation wells, arranged at several levels along the profile of the PFZ.

We have made forecasts of the spread of the brines injected into the permafrost zone by analytical methods and with simplified two-dimensional models. The next step is to create a general three-dimensional model of the process that combines hydraulic, thermomechanical, and geochemical effects.

DEPOSITS OF THE ARKHANGELSK DIAMOND-BEARING PROVINCE

The explored diamond-bearing deposits in the Arkhangelsk region are represented by steep-sloping, roughly cylindrical ore deposits approximately 200 to
400 m in diameter. A southern group of diamond pipes has been chosen for initial
industrial development.

Kimberlite deposits pierce through the Paleozoic sedimentary rock and are overlain by a 20-m to 50-m thick continuous cover of Quaternary deposits (Figure 6.3).
The poorly lithified Paleozoic stratum is represented by sandstones, argillites, and
aleurolites. Down to depths of approximately 200 m, this stratum contains a large
volume of fresh water and has a transmissivity on the order of 250 m²/day, capable
of causing inflow into the mine workings of 100,000 m³/day. At these inflow rates,
safety concerns prevented the development of the deposit by subsurface methods.
In particular, an impermeable curtain in the form of a ring of artificially frozen rock
9 km in perimeter and 200 m in depth was considered as an alternative method of
protecting the underground mine works but was rejected for both technical and

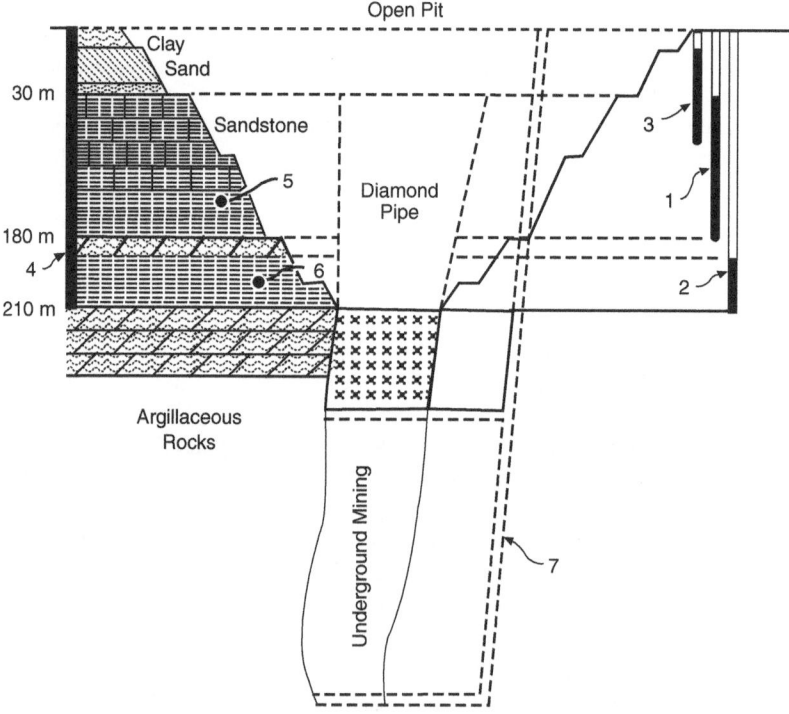

Figure 6.3 Schematic profile at the first-level development site of the Arkhangelsk deposit.
(1) Water-lowering wells at the main water-bearing complex; (2) drainage wells at
the saltwater horizon; (3) recharging wells for transfer of water from surface deposits
into the main horizon; (4) impermeable curtain made by the artificial freezing of
rock (designed alternative for underground working); (5) main horizon of fresh water;
(6) horizon of salt water; (7) designed workings of the underground mining complex.

economic reasons. In the end, it was decided to develop the deposit by the open-pit technique. The construction of a first 120-m-deep pilot-operation mine pit has now begun. A normal mining operation should be made possible by a contour system of wells to depress the groundwater levels with a total discharge of around 3000 m³/h, as well as by additional recharging wells to transfer the water of surface-deposits into the main water-bearing complex.

The proposed open pit mine could have an essential influence on the region's groundwater quality due to:

- Water-diversion facilities, including diversion of the river into a new artificial bed, tailing ponds, and accumulating ponds (Figure 6.4);
- Changes in the surface runoff due to excavation, dumping, water drainage ditches, surface regrading, etc.;
- Operation of the mine drainage and dewatering systems inside the mine pit;
- Underground recharge of some of the surface (manmade) runoff.

The end result of these complex influences may be:

- Changes in the groundwater depths, which will result in drainage of soil and marshland in areas where the level is lowered under the influence of the mine drainage, but which will result in further wetting and bogging of soil in areas of flooding near the new manmade reservoirs and water-diversion courses;
- Curtailment of the underground discharge into rivers (naturally facilitated, in particular, by the increased permeability of the enclosing rock along the contact with the ore body) and reversal of their hydraulic connections with groundwater within the zone of influence of the mine drainage—converting them from contours of discharging groundwater into areas of groundwater recharge, with corresponding reduction in runoff flow to the river;
- Reduction in usable resources of fresh groundwater in the region, as a result of drawdown by the mine-pit drainage systems;
- Pollution of groundwater and of the drainage and mine waters by flow losses from the tailings pond, as well as by infiltration of polluted water in the zones of industrial facilities, fuel and oil tanks, etc;
- Salt-water "upconing" from deep deposits to the mine and drainage workings, as well as direct inflow of salt water into the open pits during the later stage of mining;
- Pollution of the surface and groundwaters due to the surface disposal of the untreated drainage water, as well as the excessive circulating water during ore processing.

Completed studies — including a large-scale multi-well pumping test, tracer experiments, monitoring of the hydrological regime of water courses and of their connection with groundwater, and a hydrogeochemical survey — were aimed at assessing the influence of each of the above-mentioned factors on the ecological situation in the region. From these data, a permanently operating numerical model was created for flow and mass-transport in groundwater under conditions of open mining of the kimberlite pipes. From the results of the model investigation and the experimental work, the following evaluations and recommendations have been incorporated into the design decisions:

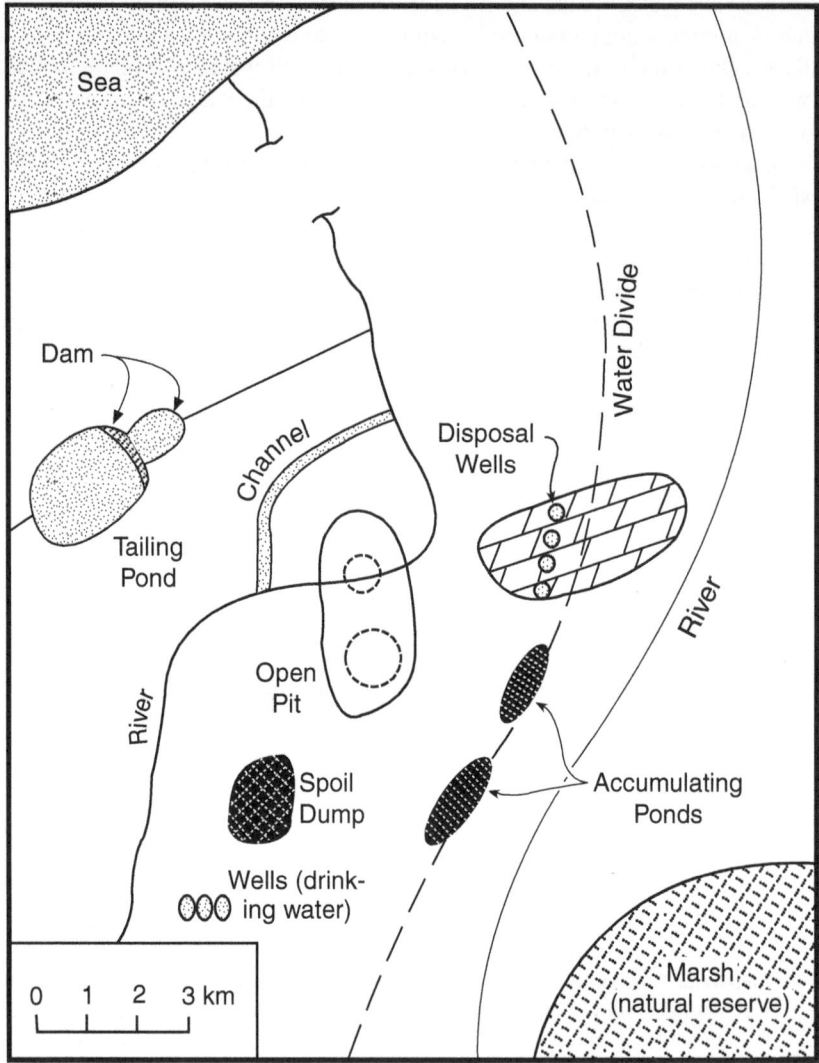

Figure 6.4 Diagram of the primary objects at the first-level working site of the Arkhangelsk deposit.

- Damage to the river runoff due to curtailment of groundwater discharge and partial reversal of their interconnection can be offset by creating a hydraulic barrier at the river's watershed (Figure 6.4) and by discharging some of the surplus treated drainage water into the river network;
- In view of the great abundance of water in the main water-bearing complex, ensuring a many-fold dilution of the infiltrating polluted waters in zones of the tailings pond and mine dumps, as well as lining the bottoms of these facilities with clays, will ensure the preservation of high quality of groundwater;

- The bulk of the drainage water pumped out by wells will be used for the drinking water supply of the city of Arkhangelsk with its half-million population; a model investigation has confirmed the reliable sanitary protection of the drainage wells as a drinking-water source;
- Harm to the usable resources of fresh groundwater could be greatly reduced by recharging water from rivers and newly created manmade basins;
- Cleaning of the contaminated mine waters before discharge into the river network will be done by filtering through peat deposits in nearby swamps;
- In view of the low capacity of the saltwater deposit lying below the main water-bearing complex, it is proposed to pump out this water by a separate system of wells and to inject it back into the overlying strata at a distance from the mine.

CONCLUSIONS

1. Northern areas are characterized by extreme hydrologic and environmental fragility. Therefore, deep mining is possible there only if specific water-protection measures, which are typically complex and expensive, are undertaken. In particular such measures are necessary in mines dug in deep permafrost when mining drainage systems pump out large volumes of brines at depth. Environmental safety of adjacent areas requires that these brines be used or stored in the subsurface.

2. The following subsurface storage options may be available for deep brines:
   - Brine-bearing strata themselves — if they have sufficient capacity and contain relatively impermeable natural barriers (i.e., subvertical faults), to prevent the disposed brines from returning into the mine workings;
   - Thick permafrost strata possessing high potential storage capacity due to pore spaces incompletely filled with water or ice. Such a storage system is indicated if the strata consists of interbedding of permeable and relatively impermeable strata, with the latter hindering the quick return of disposed brines into the drained deep brine-bearing formations. Shallow wells, therefore, are generally preferred if brines are disposed into permafrost.
   - Subsurface disposal is not indicated if high brine inflow rates (e.g., more than 1000 m^3 per hour) are anticipated. At high brine inflow rates, the mine workings may be isolated by constructing deep (on the order of hundred meters), long (on the order of kilometers) injection-type grouted barriers. Since some residual "windows" in the grouting wall are inevitable, the necessary degree of isolation (permeability less than 1 mm/day) will not likely be achieved for the entire barrier. To close the windows, the authors propose ice grouting by means of pumping fresh water into the brine-bearing strata.

4. In cold areas lacking thick permafrost zones, the main hydroecological problem is to diminish the effects of mining and drainage on the quality and quantity of nearby ground and surface waters. The most important preventive measures are:
   - The adjustment of the appropriate well-head protection zones to allow wide use of fresh drainage water for drinking supply
   - Construction of special recharge basins to limit groundwater drawdown
   - Separate pumping-out of the fresh and salt water
   - Purification of contaminated drainage water by using sorption properties of natural marsh deposits

The cost of preventive measures such as these (which are always preferable to the remediation of contaminated groundwater) may make up as much as 30 to 40% of the total cost for the mineral deposit development.

On the whole, the costs for environmental remediation and pollution prevention amount to more than a third of the projected cost of development of the deposits.

Thus, our research experience at the aforementioned sites testifies to the extreme complexity of the hydrogeoecological problems arising there and the need for qualified scientific input in the prospecting, design, construction, and mining operations. We hope that this experience will be of some interest also for mining in cold regions of the United States and Canada.

CHAPTER 7

Application of Freezing for Concentrating Solutes in Liquid Radioactive Waste Solutions

I.L. Khodakovsky, M.V. Mironenko, I.V. Chernysheva, and O.S. Pokrovsky

CONTENTS

INTRODUCTION

At the present time, a large amount of liquid radioactive wastes (LRW) has accumulated as a result of nuclear technology processing. One of the most important problems concerning the handling of LRW is the storage and purification of large volumes of low- and medium-active wastes. This study presents an investigation of the concept of concentrating radionuclides in liquid radioactive wastes (LRW) by freezing. This technology should have some advantages over other methods (e.g., evaporation, Glagolenko et al., 1996) as less ecologically dangerous and less energy-consuming, especially in cold regions.

The concentration of dissolved radionuclides in LRW is very low, ranging from 10^{-9} to 10^{-12} mol/L. As a result, aqueous radionuclides do not form a separate solid phase. Their solubility behavior is mainly controlled by coprecipitation with major salt ions (e.g., Sr with Ca^{2+} and Mg^{2+}; Cs with K^+) or sorption onto insoluble solids (Co, Ce, Zr, Nb, actinides onto iron oxyhydroxides or clays).

FREEZING OF ELECTROLYTE SOLUTIONS

Freezing-point depression is one of the most widely known solution properties. The use of freezing for LRW concentration is based on the following phenomenon: during freezing of aqueous solutions, the ice that forms on the surface (the density of ice is 0.88 to 0.92 g/cm^3) is relatively free of solutes while the bulk solution becomes concentrated in solutes. For most salts, binary electrolyte–H_2O phase diagrams have been constructed on the basis of experimental data. Such diagrams permit the determination, for a given system composition and temperature, of the phase composition, including the volumes of phases and equilibrium concentration of solution. For ternary and multicomponent systems, one can use theoretical models to estimate the composition at a given temperature (Dreving and Kalashnikov, 1964). Unfortunately, these approximations may not give exact estimates due to unpredictable interactions of various electrolytes in solution. The usual way to model electrolyte solution crystallization is a step calculation of multicomponent equilibrium compositions by free energy minimization for a multicomponent system at subzero temperatures. Such method was used by Spencer et al. (1990) for the seawater freezing modeling.

To calculate the Gibbs energy minimum of the system, the single-ion activity coefficients and osmotic coefficients of water must be known. These values may be calculated by an ion-specific interaction model; Pitzer parameters may be obtained from experimental data on solubilities in binary and ternary systems. We developed a model of equilibrium calculation at subzero temperatures combined with Pitzer procedure for activity coefficients (Mironenko et al., 1992). Unfortunately, because of the lack of information for the Pitzer parameters of solutes in LRW, we will analyze only some binary systems that are of major interest for radioactive waste freezing.

COMPOSITION OF LRW

The majority of LRW (70%) is represented by salt-rich waters (from deactivation of first contour, equipment, and tanks of biological defense); the remaining part (30%) consists of solutions with low ionic strength. Low-concentration solutions have a pH of 9 to 10, contain up to 50 mg/L of NH_3 and less than 50 mg/L of suspended particles. A typical composition of radioactive nuclides in LRW is

(a)	Co-56, Co-60	20%	(b)	Co-60	30%
	Sr-89, Sr-90	10%		Cs-134, Cs-137	20%
	Cs-134, Cs-137	10%		Sr-89, Sr-90	20%
	Ce-144, Pr-144	10%		Ce-144 + Pr-144	15%
	Ba-140 + La-144	10%		Zr-95 + Nb-95	15%
	Zr-95 + Nb-95	5%		Ba-140 + La-144	10%
	J-131	35%			

The specific activity of these waters may vary from 10^{-1} to 10^{-5} Ci/L.

Acidic brines have an activity ranging from 10^{-5} to 10^{-2} Ci/L; examples of these solutions are the following:

(a) pH = 1.6 to 2.3, $[H_2C_2O_4]$ = 0.02 M, salts of Fe^{3+}, Mn^{2+}, and Ni^{2+} may consist up to 0.01 M

(b) pH = 1.5, $[H_2C_2O_4]$ = 0.02 M, $[H_3PO_4]$ = 0.02 M, $[H_2O_2]$ = 0.15 M, $[Mn^{2+}]$ = 0.012 M, $[Fe^{3+}]$ = 0.005 M, $[Cr^{3+}]$ = 0.0004 M

(c) pH = 2.5, $[HNO_3]$ = 0.5 M, $[H_2C_2O_4]$ = 0.03 M, $[Mn^{2+}]$ = 0.013 M, $[H_3PO_4]$ = 0.0004 M

(d) $[H_2C_2O_4]$ = 0.03 M, $[(NaPO_3)_6]$ = 0.005 M, 1.5 g/L of sulfanol

Some examples of LRW composition may be found in Bradley (1997).

IONIC EXCLUSION FROM FREEZING SOLUTIONS

Equilibrium Model: Ideal Freezing

Factors controlling the distribution of dissolved substances (including radionu-clides) between the ice and residual solution are of critical importance in under-standing practical problems associated with LRW concentration by freezing. The equilibrium distribution coefficient for solute (K_o) during crystallization of electro-lyte solution is described by the equation:

$$K_o = N_s/N_1 = 1 - \Delta H \cdot \Delta T/(RTm^2 \cdot N_1) \qquad (7.1)$$

where N is the mole fraction of dissolved component in the ice (N_s) and in the solution (N_1), T_m is the freezing temperature of pure solvent (K), ΔT is the freezing point depression (K), and ΔH is the heat of fusion of solvent (79.69 cal/g for H_2O). According to Leung and Carmichael (1984), ideal freezing of the equilibrium dis-tribution coefficient of solute (K_o) may be found from initial conditions of freezing:

$$C_s/C_1 = K_o \cdot (1-g)^{K_o-1} \qquad (7.2)$$

where C_s and C_1 is the current concentration of solute in solid and liquid, respectively, and g is the fraction of total volume frozen.

The modeling of initial solution alteration during freezing was carried out based on the solubility diagrams of binary systems: H_2O-HNO_3, H_2O-KOH, H_2O-

$K_2C_2O_4$. We assumed that the density of solution was 1 g/cm³ and initial volume was 1 m³. The system H_2O-HNO_3 with initial HNO_3 concentration of 30 g/L corresponds to the type c of acidic LRW. The final concentration of HNO_3 (at eutectic point) reaches 32%. We determined the concentration factor (k) as a ratio of initial concentration of solute in liquid (before freezing) to the final concentration of solute in remained liquid corresponding to eutectic composition. For different initial composition, the concentration factor in eutectic may vary from 3 to 11. For the system H_2O-KOH with initial KOH concentration of 6.5 g/L, the temperature of eutectic is –60.9°C and concentration factor is 30 to 40 (Figures 7.1, 7.2, and 7.3). The eutectic phase compositions and eutectic temperatures of several water-salt systems are listed in Table 7.1, and the results of calculation of solution concentrations at subzero temperatures are given in Table 7.2. As can be seen, the predicted equilibrium distribution coefficients of salt between ice and remaining solution is high enough. Assuming that the behavior of trace amounts of dissolved radionuclides is similar to that of major ions, it is possible to use the calculated concentration factors of solutes for rough estimation of the radionuclides distribution in the process of LRW freezing.

Incorporation of Impurities into the Ice during Nonideal Freezing

Usually we deal with nonideal freezing of electrolyte solutions. In multicomponent solutions, continued growth of the solid phase depends not only upon the ability to remove the latent heat of fusion, as in the case for a pure substance when no solutes are present, but also upon the ability of the solutes in the liquid phase to

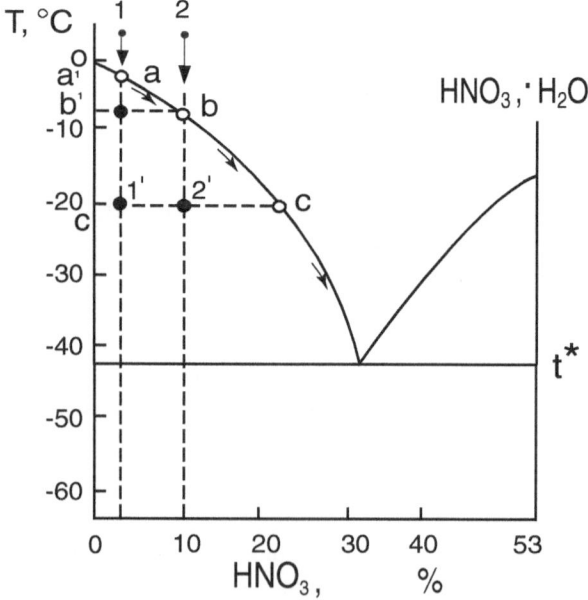

Figure 7.1 Phase diagram of the system HNO_3-H_2O.

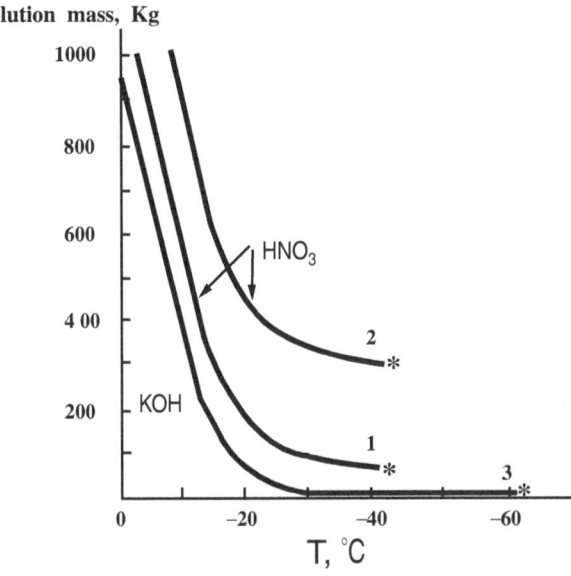

Figure 7.2 Change of masses of remaining solutions in the freezing process. (1) HNO_3, $C_{initial} = 3\%$; (2) HNO_3, $C_{initial} = 10\%$; (3) KOH, $C_{initial} = 0.65\%$; * = eutectic points.

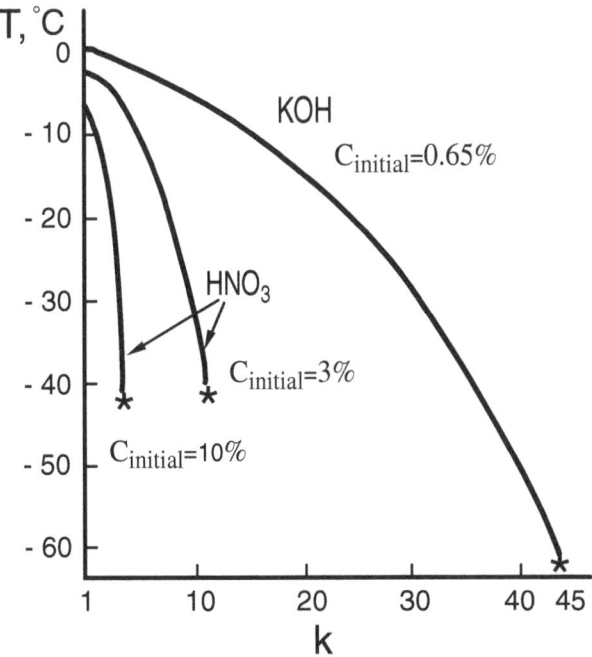

Figure 7.3 Concentration coefficient as a function of temperature.

Table 7.1 Eutectic Phase Compositions and Eutectic Temperatures of Some Water-Salt Systems

System	Eutectic Temperature, °C	Phase Composition of the Eutectic
$CaBr_2$-H_2O	−83.0	$CaBr_2$ + ice
$CaCl_2$-$MgCl_2$-H_2O	−52.2	$CaCl_2 \cdot 6H_2O$ + $MgCl_2 \cdot 12H_2O$ + ice
$CaCl_2$-H_2O	−49.8	$CaCl_2 \cdot 6H_2O$ + ice
$MgCl_2$-H_2O	−35.0	$MgCl_2 \cdot 12H_2O$ + ice
Na_2CO_3-K_2CO_3-H_2O	−37.0	$(Na,K)_2CO_3 \cdot 6H_2O$ +ice
NaCl-KCl-H_2O	−23.5	$NaCl \cdot 2H_2O$+KCl + ice
KF-H_2O	−21.5	$KF \cdot 4H_2O$ + ice
NaCl-$NaHCO_3$-H_2O	−21.8	$NaCl \cdot 2H_2O$ + $NaHCO_3$ + ice
NaCl-Na_2CO_3-H_2O	−21.4	$NaCl \cdot 2H_2O$ + Na_2CO_3 + ice
KCl - H_2O	−10.6	KCl + ice $Na_2SO_4 \cdot 10H_2O$ + ice
Na_2SO_4-H_2O	−1.2	

Table 7.2 The Results of Calculation of Concentration Coefficients during Freezing of Solutions

System	H_2O-HNO_3	H_2O-HNO_3	H_2O-KOH	H_2O-$K_2C_2O_4$
Initial content, of solute, %	3	10	0.65	3.5
$T_{crystallization}$, °C	−2.3	−6.7	0	0
T = −3°C				
$m_{ice}/m_{solution}$	—	—	—	740/260
k	—	—	—	3.8
T = −20°C				
$m_{ice}/m_{solution}$	877/123	570/430	960/40	—
k	8.1	2.3	25	—
T = −30°C				
$m_{ice}/m_{solution}$	889/111	666/334	969/31	—
k	9	2.9	32	—
T = −40°C				
$m_{ice}/m_{solution}$	909/91	678/323	972/28	—
k	11	3.1	36	—
T = −60°C				
$m_{ice}/m_{solution}$	—	—	977/23	—
k	—	—	43	—
$T_{eutectic}$, °C	−42	−42	−60.9	−5.88
Final content of solute, %	32	32	29.3	19.4

diffuse away from the interface, thereby permitting the solvent to gain access to the solid phase. In that case, to calculate the distribution of solute between liquid phase and ice, the so-called Burton-Slighter approach may be used (Leung and Carmichael,

1984). In this model, the behavior of solute is analyzed in the diffuse boundary layer on the surface of growing solid phase. The transport equation should be written as:

$$K = K_o/\{K_o + (1 - K_o) \cdot \exp(-\Delta)\} \tag{7.3}$$

where K is the effective distribution coefficient, K_o is the equilibrium distribution coefficient (Equation 7.1), $\Delta = \delta f/D$, where f is the freezing rate (mm/s), δ is the thickness of diffuse nernstian layer (mm), and Δ is the molecular diffusivity (mm/s^2). According to Terwilliger and Dizio (1970), $\delta = (50-70) \times 10^{-3}$ mm at $f = 40 \times 10^{-3}$ mm/s and $\delta = 0.5-0.7$ mm when $f = 4 \times 10^{-3}$ mm/s. For many solutes in water the D value is close to 10^{-4} mm/s^2.

During the freezing of aqueous solution, an important phenomenon may be the diffusional supercooling, when the actual temperature of solution is lower than T_{eq} (equilibrium temperature of freezing) due to difference in values of heat conductance and molecular diffusion. This leads to the alteration of boundary surface: the ice dendrites grow (Harrison, 1965) following the abrupt increase of distribution coefficient. Such a "physical incorporation" of solute pockets in growing ice crystals will obviously diminish the effectivity of solute concentration by freezing. The physical incorporation of dissolved substances into the ice structure (bulk entrapment) may be eliminated by slowing down the cooling rate to achieve close-to-equilibrium distribution ratios.

Another type of incorporation is the structural (chemical) one, when the molecules of solute substitute H$^+$ or OH$^-$ ions in ice structure in different ways. It is especially important for cations isomorphic with H$^+$ (e.g., NH$_4^+$) or anions isomorphic with OH$^-$ (e.g., F$^-$, Cl$^-$). According to Rozental (1973) the only known substance that can form a solid solution in ice is NH$_4$F. The degree of structural incorporation may be estimated from the value of freezing potential. It is an electric potential that is measured between the liquid phase and the ice by means of inserted electrodes. This potential is proportional to the amount of ions incorporated into the ice and coincides with the sign of ions; i.e., it is negative in case of preferential capture of anions and positive for cations. The freezing potentials were measured for many salts (Rozental, 1971) and they are listed in Table 7.3. Although the absolute value of freezing potential does not allow us to predict directly the distribution coefficient of trace component between the ice and the liquid phase, it provides a rough estimation of the degree of ion incorporation into the ice structure during freezing. For example, as can be seen in Table 7.3, the important components of LRW, ammonia, nitrate, bicarbonate, and oxalate are likely to incorporate into the solid phase. In contrast, we should not expect structural entrapment of large radioactive cations (e.g., Sr^{2+}, Cs^{2+}, Co^{2+} as well as transuranium elements and their complexes in aqueous solution), which should remain in the liquid phase during freezing. Maximal values of freezing potentials usually may be achieved at salt concentration of 0.0002 to 0.00002 M. It means that the chemical incorporation of impurities into the ice structure during freezing is more important at low concentration of salts than in the concentrated solutions (Rozental, 1968, 1971, 1972, 1973).

In Table 7.4 we compiled the existing literature data on the solute distribution during freezing of electrolyte solutions that are discussed below. Leung and

Table 7.3   Electric Potentials of Freezing for Different Electrolyte Solutions

Salt	E, mV	Salt	E, mV
NH_4F	−9	KI	−25
KF	−40	$(NH_4)_2CO_3$	180
NaF	−46	Na_2CO_3	−70
NH_4Cl	92	$KHCO_3$	−41
KCl	−37	$NaHCO_3$	−55
NaCl	−43	$(NH_4)_2 C_2O_4$	84
KNO_3	−31	$K_2C_2O_4$	−28
$NaNO_3$	−27	$Na_2C_2O_4$	−31

Compiled from Rozental, 1968, 1972, 1973.

Carmichael (1984) found a good agreement between measured and predicted equilibrium values, so at their freezing rate the distribution of solute approached the ideal one. Saidov et al. (1990) investigated the freezing of river water and found K values from 0.1 to 0.2 for the major dissolved cations. Romanov and Levchenko (1989) measured $K = 0.28$ to 0.35 for pure NaCl. A thorough investigation of salt rejection phenomena in the freezing of salt solutions, including the behavior of trace components, was performed by Chepurnaya (1975). According to her results (Table 7.4) the enrichment of remaining solution is very high,

Table 7.4   Distribution of Solute during Freezing of Electrolyte Solutions

Solute	Freezing Rate (f)	$K_{MEASURED}$	K_o	Source
$CaCl_2$	12.5 mm/h	0.54	0.52	Leung and Carmichael
$BaCl_2$		0.60	0.51	(1984)
HCO_3^-	Not	0.26		Saidov et al. (1990)
Cl^-	reported	0.36		
SO_4^{2-}		0.22		
Ca^{2+}		0.30		
Mg^{2+}		0.49		
$Na^+ + K^+$		0.13		
NaCl	1.6 mm/h	0.31 ± 0.04		Romanov and Levchenko (1989)
SO_4^{2-} in KCl	7.5 mm/h	0.41 ± 0.03	0.28	Chepurnaya (1975)
SO_4^{2-} in KBr	7.5 mm/h	0.47 ± 0.06	0.35	
SO_4^{2-} in CsJ	7.5 mm/h	0.10	0.03	
Cu^{2+} in CsJ	7.5 mm/h	0.089 ± 0.017		
Fe^{2+} in CsJ	7.5 mm/h	0.025 ± 0.007		
Rb^+ in KNO_3	7.5 mm/h	0.35 ± 0.02		
Sr-90	Not	0.0028		Baturin et al. (1978)
Cs-137	reported	0.0015		
Co-60		0.0031		
Ce-144		0.010		
Ru-106		0.100		

e.g., the concentration factor of Rb^+(aq) in KNO_3 solution achieved 50 to 60. The most important conclusion of this study was that the major salts and not the impurities play the main role in separation of solute during the freezing of electrolyte solutions. This allows the using of basic well-characterized salt–H_2O systems ($KCl–H_2O$, $HNO_3–H_2O$, $H_2C_2O_4–H_2O$, etc.) for quantitative description of the freezing of electrolyte solutions corresponding to LRW.

There are much less data concerning the distribution of radioactive elements between aqueous solution and ice. Baturin et al. (1978) studied the freezing of natural water reservoir during winter and reported the distribution of several radionuclides. The average distribution coefficients are very low (Table 7.4) demonstrating a high purification degree of aqueous solutions in the processes of natural freezing. Smagin (1997) reported the results of Sr-90 and Cs-137 distribution between ice and water obtained on one reservoir of Production Association "Mayak" (southern Urals, Russia). The freezing of this artificial lake during winter results in formation of a 50-cm-thick ice layer where the concentrations of Sr and Cs depend on the depth profile. In solution of 0.004 M $MgCl_2$, 0.007 M $CaCl_2$, and 0.015 M Na_2SO_4, the distribution coefficients of Sr and Cs were in the range of 0.5 to 0.04 and 0.004 to 0.07, respectively. Unfortunately, lack of information concerning rate of freezing and chemical composition of initial solutions allows the using of these data (Baturin et al., 1978; Smagin, 1997) only as qualitative estimation of the radionuclide's behavior at subzero temperatures.

Growth Rate of Ice Crystals in Aqueous Solutions

Solutes in a concentration of 0.01 M produce a slight increase in the ice growth rate, while at greater concentrations the growth rate decreases. According to Lindenmayer and Chalmers (1966), the growth rate (v, mm/s) in KCl solution at a concentration of less than 0.001 M is the same as that in pure water. It is described by the following equation:

$$v = 0.228 \cdot \Delta t^{2.39} \text{ mm/s}$$

where Δt is the supercooling of solution, i.e., the difference between actual temperature of freezing and that at equilibrium. The growth rate decreases at higher concentrations of salt:

$$v = 0.142 \cdot \Delta t^{2.04} \text{ in 1 M KCl}$$

$$v = 0.112 \cdot \Delta t^{2.06} \text{ in 1 M CH}_3\text{COOH}$$

The upper limit of supercooling (and corresponding growth rate) for LRW crystallization is 5 to 7°C. Above this value, the spontaneous nucleation of ice may occur (Lindenmayer and Chalmers, 1966). That may lead to very high growth rates and physical incorporation of solution into the growing ice, including formation of solute pockets and bulk entrapment of the liquid. As a result, the concentration factor of the solute will decrease.

GENERAL RECOMMENDATIONS FOR USING FREEZING FOR CONCENTRATING LRW

Analysis of solubility diagrams of binary systems representing the major types of LRW solutions demonstrates that in the case of equilibrium freezing the concentration factor reaches 30 to 40. Thus the freezing of LRW is able to concentrate the solute (including minor radioactive components) by 30 to 40 times. The key point is the duration of cooling. Below the eutectic (cotectic for multicomponent systems) point the co-crystallization of ice and salt crystallohydrate occurs. Thus, it is necessary to stop the freezing before reaching the cotectic point. The slower the rate of cooling, the fewer solutes will be incorporated in the ice so one can decrease the distribution coefficient K of radionuclide by slowing down the process of freezing and achievement of the equilibrium distribution. The remaining radioactive solution, corresponding to the eutectic point, may be disposed using the usual procedures of handling radioactive wastes. It is also important to eliminate the effect of nonideal freezing, e.g., structural incorporation of solute and physical entrapment of solution into the solid phase. The freezing should be conducted at the conditions close enough to the equilibrium. The optimal value of supercooling would be 2 to 3°C when the rate of ice growth is still high enough, whereas the bulk entrapment of solute pockets is unlikely.

For various compositions of LRW considered, the freezing is possible at conditions of permanently frozen soils (–7°C, outside temperature in the winter down to –40°C in northern areas) as well as with the use of cryogenic techniques.

ACKNOWLEDGMENTS

The authors are grateful to Steven A. Grant, who offered many thoughtful comments, made many editorial suggestions, and provided useful literature references. Financial support of the Russian Foundation of Basic Research (Project Number 97-05-64197) and the Research, Development and Standardization Group (U.K.), U.S. Army Materiel Command (Project No. R&D 7807-EN-06), is acknowledged.

REFERENCES

Baturin, V.A., I.G.Vodovozova, and L.N. Korchak. On the Distribution of Some Long-Lived Radioactive Elements in Water Reservoir and Their Transfer into the Ice. *UFAN SSSR (Sverdlovsk)*. 114, 70–73, 1978.

Bradley, D.J. *Behind the Nuclear Curtain: Radioactive Waste Managements in the Former Soviet Union*, Payson, D.R., Ed., Battelle Press, Columbus, Ohio, 273–274, 1997.

Chepurnaya, V.G. The Normal Crystallization of Salt Solutions as a Method of Concentration of Microimpurities (in Russian), thesis presented to the University of Novosibirsk, Russia, in fulfillment of the requirements for the degree of Doctor of Philosophy, 1975.

Dreving, V.P. and Ya.A. Kalashnikov. *The Phase Rule*. Moscow University Press, Moscow, 1964.

Glagolenko, Yu.V., E.G. Dzekin, and E.G. Drozhko. Handling of Radioactive Wastes on Production Association "Mayak." *Radiat. Saf. Prob.*, 2, 3–10, 1996.

Harrison, J.D. Solute Transpiration Pores in Ice. *J. Appl. Phys.* 36, 1–15, 1965.

Leung, W.K.S. and G.R. Carmichael. Solute Redistribution During Normal Freezing. *Water Air Soil Poll.* 21, 141–150, 1984.

Lindenmeyer, C.S. and B. Chalmers, Growth Rate of Ice Pendrites in Aqueous Solutions, *J. Phys. Chem.* 45, 2807–2808, 1966.

Mironenko, M.V., M.Yu. Zolotov, and M.Ya. Frenkel. The Algorithm, Program Code and Database for Computation of Chemical Equilibrium in Multicomponent Systems, Containing Non Ideal Aqueous, Gas and Solid Solutions, in *Thermodynamics of Natural Processes '92. Proceedings of the Second Intern. Symposium*, September 13–20, 1992, Novosibirsk, Russia. Siberian Branch of RAS, Novosibirsk, 1992, 112.

Romanov, V.P. and G.P. Levchenko. Migration of Salts in Freezing Soils. *Sov. Eng. Geol.* 2, 44–51, 1989.

Rozental, O.M. The Structural Features of Water and Electrokinetic Effect of Its Crystallization. *Russ. J. Struct. Chem.* 9, 777–779, 1968.

Rozental, O.M. The Hydration Complexes of Ions: Their Structure and Incorporation into the Ice During Freezing. *Russ. J. Struct. Chem.* 12, 917–919, 1971.

Rozental, O.M. The Crystallization of Ice in Water and Aqueous Solutions. II. Kinetics of Crystallization of Aqueous Electrolyte Solutions. *Russ. J. Phys. Chem.* 46, 657–659, 1972.

Rozental, O.M. The Connection Between Properties of Ionic Solution and Ice Formation Processes. *Russ. J. Struct. Chem.* 14, 797–801, 1973.

Saidov, M.S., A. Avezmuratov, and E.A. Koshchanov. The Interaction and Distribution of Impurities in Two-Phase System Aqueous Solution — Ice. *Dokl. Uzbek. Academ. Nauk.* (in Russian). 2, 30–31, 1990.

Smagin, A.I. Distribution of Radionuclides in the Ecosystem of Waste Storage Reservoir and the Estimation of the Effectivity of Water Deactivation by Freezing. *Rad. Saf. Prob.* 1, 64–70, 1997.

Spencer, R.J., N. Møller, and J.H. Weare. The Prediction of Mineral Solubilities in Natural Waters: A Chemical Equilibrium Model for Na-K-Ca-Mg-Cl-SO_4-H_2O System at Temperatures below 25°C. *Geochim. Cosmochim. Acta*. 54, 575–590, 1990.

Terwilliger, J.P. and S.F. Dizio. Salt Rejection Phenomena in the Freezing of Saline Solutions. *Chem. Eng. Sci.* 25, 1331, 1976.

CHAPTER **8**

Feasibility and Computer Model Studies of Isolation of Radioactive Wastes in Perennially Frozen Bedrock

A.N. Kazakov, N.F. Lobanov, M.V. Mironenko, A.I. Shapkin, and S.A. Grant

CONTENTS

INTRODUCTION

The All Russian Research and Design Institute of Production Engineering (VNIPIProm-technologii Institute) of the Ministry of Atomic Energy (MINATOM) of the Russian Federation is studying burial in perennially frozen bedrock on the Novaya Zemlya archipelago as a radioactive-waste isolation technology (Figure 8.1). The radioactive waste (RAW) is expected to come from spent nuclear reactor fuel from submarines and icebreakers. The current development of this large-scale demonstration is being conducted within the Russian Federal Target program entitled "Management of Radioactive Wastes and Spent Nuclear Materials, Their Utilization and Disposal in 1996–2005." In principle, this is a potential topic for research collaboration between U.S. Army Cold Regions Research and Engineering Laboratory (CRREL) and VNIPIPromtechnologii Institute. Because of limited funding for this topic, no substantive collaboration has occurred.

Frozen water-saturated porous media have extremely low permeabilities and great mechanical strengths. Due to these properties, artificially frozen ground has long been used to reduce seepage into and stabilize walls of soil excavations (Harris, 1995). These properties have been exploited more recently for isolating contaminated soils with barriers formed from artificially frozen ground. Since rock is generally stronger and less permeable than soil, it is natural to consider perennially frozen bedrock as waste isolation media. Since the permafrost depths in Novaya Zemlya

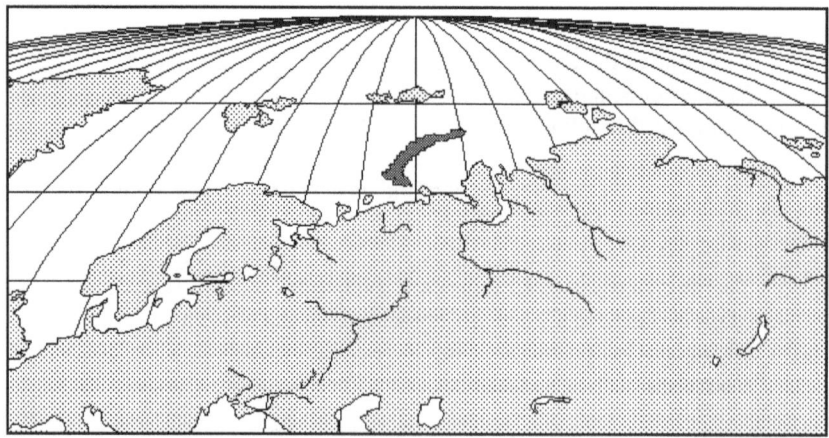

Figure 8.1 Map of northern Europe showing the location of Novaya Zemlya archipelago.

are great (over 300 m), if energy fluxes from the individual RAW containers are not too large, it is likely that permafrost would be able to isolate the waste thermally as well as mechanically and hydrologically.

FEASIBILITY STUDY

Design Concept

Provided that the permafrost is thick enough and that the heat generated by disposed RAW is small relative to the annual energy balance at a site (i.e., will not melt an unacceptably large volume of permafrost), permafrost could be an attractive repository for RAW disposal.

Site Selection Criteria

Three criteria have been developed to evaluate candidate radioactive-waste isolation sites in permafrost:

1. The permafrost should not contain any free liquid water, for example, as interpermafrost aquifers or cryopegs.
2. The permafrost should be thick enough to isolate the RAW thermally, radiologically, and hydrologically.
3. The net energy loss in winter should be sufficient to prevent appreciable thawing due to the energy generated by radioactive decay in the waste containers.

Storage Repository Designs

Based on current practices for mining and drilling in permafrost, four repository designs have been developed:

1. Deepened trenches — for isolation of low-level or intermediate-level RAW
2. Shafts or boreholes — for isolation of intermediate-level RAW and high-level RAW
3. Boreholes — for isolation of high-level RAW
4. Adits of various cross sections — for RAW isolation

Site Description

In 1991 VNIPIPromtechnologii Institute conducted the feasibility study (FS) for construction of a large-scale demonstration for underground storage of RAW generated by the Northern Fleet of the Russian Navy (principally from spent nuclear reactor fuel from submarines and icebreakers).

The site of the proposed demonstration is on the southwestern end of the Southern Island, 15 km to the northeast of the pier in the Bashmaclmaya Bay (Figure 8.2). The proposed demonstration site is situated on a coastal plain terrace with elevations ranging from 50 to 140 m, having many swampy areas and numerous streams and

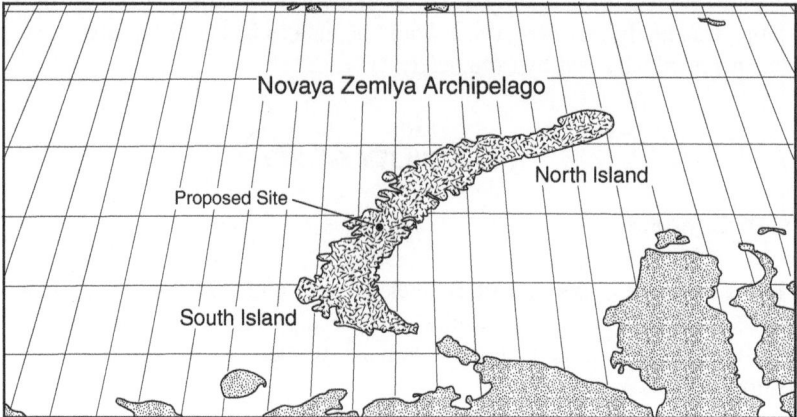

Figure 8.2 Map of the Novaya Zemlya archipelago showing the location of the proposed site.

lakes. The site is a flat hilltop (Mt. Chernaya) with elevations ranging from 100 to 138.5 m. While a route to the site was surveyed in 1993, one has not been constructed and the site is accessible only by tracked vehicles.

The geomorphology is typical of cryogenic microrelief, with pingos, stone polygons, and soil-flow terraces. The vegetation of the general area in and around the site is forest-free tundra, consisting of moss and either bushes or lichen.

Preliminary geological surveys of the site have indicated that it lies on limestone of Upper Devonian and Lower Carbonic age interbedded with slates and sandstone. The Upper Devonian limestones are interstratified with slates, weakly sandy limestone, and aleurolite. Lower Carbonic limestones are organic, sandy in places, nonuniformly dolomitic, brecciated in local places, and contain fine grains of pyrite. Rocks are strongly dislocated, crushed in abrupt folds at the northwest strike.

The rocks near the surface are fractured. To depths of 10 to 30 m, the fractures are filled by ice and by minerals — mainly calc-spar, less often quartz, sometimes pyrite, or iron hydroxides. The fractured bedrock is overlain discontinuously by a 1- to 2-m-thick bed of loose Quaternary deposits generally having a loam or loamy coarse sand texture.

The temperatures at the base of the Quaternary deposits vary annually from 3.5 to –4.5°C. Permafrost apparently extends to depths of 300 to 350 m. The active layer is 0.5 to 2.0 m thick and is thawed from June to October. The active layer is generally unsaturated, but seepage could be appreciable into ditches and building foundations.

Feasibility Study Conclusions

In 1992 state regulators approved the feasibility study, which supported the concept and tentative designs for and siting of RAW storage in permafrost. The feasibility study concluded the following:

1. The geological and geophysical surveys on the Novaya Zemlya archipelago indicate that the geomorphology, stratigraphy, and nature of permafrost on this island meet the site selection criteria for an underground RAW repository.
2. Novaya Zemlya is isolated from other industries and populations and itself has no native population, industry, or agriculture, indicating that the siting may be intrinsically safe.
3. Because of the underground nuclear weapons testing that was conducted on Novaya Zemlya, there is substantial expertise on drilling and excavating in permafrost and on handling radioactive materials in cold environments.
4. No competing future economic development is contemplated for the proposed RAW repository, which would be built on the territory of the former nuclear test site.
5. The region is currently under federal control. Were this to change in the future, it is likely that the federally established policies, standards, and practices would be largely retained.

It is considered an advantage that the proposed repository is on an island, possibly limiting overland transportation of RAW.

MODELING STUDIES

Two modeling studies were conducted of the geophysics and geochemistry of the proposed burial system.

Geophysical Modeling

Problem Statement

Describing the convective migration with groundwaters damped by the temporary heating of rocks is central to mathematical modeling of energy, water, and solute transport near RAW containers placed in perennially frozen bedrock. In previous studies, molecular diffusion was considered to be the sole radioactive contaminant-transport mechanism for intact containers. Should the container be breached (by corrosion, for example), water, which circulates by convection in the thawed zone, would come into contact with RAW. Thus, the worst-case scenario would find all of the container contents released to the partially thawed zone around the container. A pore-scale description of the physics and chemistry in the partially thawed zone is essential. It is necessary to know the maximal volume of the partially thawed region, its time maximum, and its duration as functions of thermal-physical characteristics of the burial and the medium. In practice, it is important to determine conditions under which the permafrost resists thawing. Obviously, radioactivity could be dispersed more widely should the water in the rock surrounding the container vaporize. Vaporization would increase the interporous pressure, forming new — and expanding existing — rock fractures. Therefore, obtaining the vaporization conditions is also necessary, though these conditions were simulated in this study. The greatest energy production studied, 500 W m^{-3}, resulted in predicted temperatures of less than 80°C at the container's surface.

Modeling Assumptions

Geometry of the buried container was assumed to be a two-dimensional problem with radial symmetry at each horizon. Two coordinates determined each point of the medium: the depth h (m) and the distance from the axis of symmetry r (m), which was the center line of a cylindrical container. Eschewing a spherical-symmetrical approach allows convective water flows and chemical interactions to be modeled realistically.

The working volume of the burial was considered to be of finite size, but small enough relative to the depth of the burial. The working volume of the burial shape (a sphere, a cylinder, and so on) is the parameter of the model. There are no special assumptions about the shape. The initial temperature at the container's surface is $T_{sur}(t = 0 \text{ s}) = T_0$ (K). The working volume of the burial is V (m^3); the integral power of heat sources inside the volume is $W(t)$ (kW m^{-3}).

We modeled the rock mass as a system of nested zones:

Zone 1 is the rock mass area, immediately surrounding the working volume of the burial; the ice was entirely melted in pores of rock mass.

Zone 2 is the rock mass area, in pores of which both water (a solution of salt, in general case) and ice exist.

Zone 3 is the rock mass area, the temperature of which is more than dT higher than the temperature in the point at the initial time moment.

These zones are presented schematically in Figure 8.1.

The exterior boundary of zone 3 determines the area of the temperature wave propagation from the source of the thermal excitation, i.e., the working volume of the burial. The parameter of the model characterizes the boundary of the unfrozen zone.

We analyzed the mutual influence of the temperature on the free surface and the temperature wave from the heat source. We considered temperature oscillations at the surface, taking Tikhonov and Samarskii (1990) as a starting point. We concluded that if the heat source power $W(t)$ and the burial depth L provide the existence of the boundary of zone 3, which is situated at a depth more than 10 m in any time, then season temperature oscillations at the surface do not influence the modeling of the analyzed system.

The porosity $e(h, r, t)$ of the medium is assumed to be the function of the process of chemical interaction of the salt solution with the rock matrix.

We followed a traditional approach to the mathematical description of the thermal problem with phase transitions applying the generalized heat capacity problem (Stefan problem).

Constitutive Equations

Three constitutive equations were developed:

1. The equation of heat transfer, which expresses the law of thermodynamic potential (energy) conservation. This law usually postulates conditions under which Fourier's law can be applied:

$$\rho(t, x)T\left[\frac{\partial S(t, x)}{\partial t} + v \cdot \nabla S(t, x)\right] = \nabla \cdot [k(t, x)\nabla T]\qquad(8.1)$$

where $S(t, x)$ is the of entropy function at point x at time t (J K^{-1} kg^{-1}); k, thermal conductivity (W m^{-1} K^{-1}); $\rho(t, x)$, medium substance density (kg m^{-3}); T, the temperature (K); v, the vector of flow velocity (m s^{-1}) (Bird et al., 1960, p. 350).

2. The equation of medium continuity:

$$\frac{\partial \rho(t, x)}{\partial t} + \nabla[\rho(t, x)v] = 0\qquad(8.2)$$

3. The equation of substance state, which correlates thermodynamic and mechanical characteristics of the medium: $\rho = \rho(p, T)$; $S = S(p, T)$, where p is pressure (Pa).

Formulating the problem to solve for (S, t) is advantageous. The temperature is thus a function of entropy rather than following the traditional approach of solving $T(t)$ because system phase compositions are uniquely defined at phase transitions not by temperature, but by specific enthalpy $H(T, t)$ (J kg^{-1}). If Gibbs-energy changes across phase boundaries are assumed to be zero, enthalpy is uniquely specified by the entropy.

Let us consider that $H(t = 0, x) = 0$, $T(t = 0, x) = T_0(x)$, $T(t, x = \infty) = T_\infty$.

We should formulate the problem in terms of enthalpy $H(T, x)$. In so doing, we assume that the volume containing the buried container is discretized by cylindrical network of volume elements with height increments h_i and radius increments r_j ($i = 1, \ldots N_h$; $j = 1, \ldots N_r$).

Here the container is placed in the element ($h_k = h = h_{k+1}$, $0 = r = r_1$).

The surface square of the (i, j) element can be found by formula ($r_0 = 0$):

$$\sigma_{ij} = 2\pi - (r_{j+1} + r_j)h_i + 2\pi(r_{j+1}^2 - r_j^2)$$

Basing on Fourier's law $J_q = -k(dT/dl)$, where J_q is heat flux (W m^{-2}), we can calculate the enthalpy, which crosses each of four volume element (i, j) boundaries in time τ_n:

$$H_{ij}^{(1)} = -[\tau_n(k_{i,j} + k_{i,j+1})/2]\{(T_{i,j+1} - T_{i,j})/[(r_{i,j} + r_{i,j+1})/4\pi]\}\qquad(8.3)$$

for all admissible k(m).

If the element (i, j) contains the container then its heat content increases by $\tau_n W(t)V$, where $W(t)$ is a specific (volumetric) power of the source (W m^{-3}), and V is the part of the working volume, which belongs to the volume element (i, j).

Method of Solution

The computing program MELTPOR (FORTRAN 77) was adapted to mathematically model the process of the thermal medium evolution (i.e., the dynamics of the melting of the pore ice ignoring material transport) (Shapkin, 1992, 1993). Initial parameters of the program are presented in Table 8.1. Calculations were made in the time interval of 0 to 2 years (about 0.5 hour of processing time), so the effect of the heat source power variation was insignificant. The predicted evolution of temperature at the ground surface is presented in Figure 8.3. The predicted volume of unfrozen rock surrounding the container is presented in Figure 8.4.

Geochemical Modeling

Objective

The first overall objective of the project was to assemble the fundamental information for modeling chemical interactions that would occur in the permafrost rock surrounding a RAW container. This first overall objective included three tasks:

1. Choosing, collecting, and processing the necessary thermodynamic data for simulation of the fate and transport of actinides in permafrost rock.
2. Calculating principal chemical interactions taking place in various parts of the repository: (1) between the steel container and the pore water; (2) between the steel container, spent nuclear fuel, and water solutions; and (3) between pore water solutions (which may contain radionuclide species) and the limestone surrounding the steel container. These calculations are based on the calculated mass and energy transfers and the chemical composition of surrounding rocks and chemical composition of the spent nuclear fuel. These calculations should let us estimate the main interactions responsible for the radionuclide mobilization or immobilization. For the initial study, we selected two actinides for further examination: uranium and plutonium.
3. Developing principles for coupling the chemical interaction model with the mass-and-energy transfer submodel.

Case History Data Needed for Chemical Interaction Modeling

Limestone is the dominant rock at the proposed repository. Mean chemical composition of this material is shown in Table 8.2. The chemical analysis is consistent with an oxidized redox environment. Chemical interactions simulating this chemical composition have been calculated with the assumptions that the carbon dioxide is lost on heating, iron is present as Fe II, and sulfur is contained in pyrite.

Some physical parameters of the rock are as follows:
Effective volumetric porosity   4%
Filtration coefficient $0.16 \text{ m} \cdot \text{day}^{-1}$
Heat capacity $920 \text{ J} \cdot \text{K}^{-1} \cdot \text{kg}^{-1}$
Thermal conductivity $2.9 \text{ W} \cdot \text{m}^{-1} \cdot \text{K}^{-1}$

Table 8.1 Symbols, Quantities, and Assigned Values for Model Variables

Symbol	Quantity	Dimension	Assigned Value
ρ_r	Specific density of rock matrix	kg m^{-3}	2,800
ρ_w	Liquid solution density	kg m^{-3}	1,000
ρ_i	Density of the porous ice	kg m^{-3}	920
ρ_c	Specific density of container	kg m^{-3}	4,920
C_r	Heat capacity of rock matrix	J K^{-1} kg^{-1}	920
C_w	Heat capacity of liquid water	J K^{-1} kg^{-1}	4,200
C_i	Heat capacity of ice	J K^{-1} kg^{-1}	1,933
C_c	Container heat capacity	J K^{-1} kg^{-1}	920
C_k	Container content's heat capacity	J K^{-1} kg^{-1}	2.9
k_r	Thermal conductivity of rock	W m^{-1} K^{-1}	2.9
k_w	Thermal conductivity of water	W m^{-1} K^{-1}	0.6
k_i	Thermal conductivity of ice	W m^{-1} K^{-1}	2.23
W	Volume power of the heat generator	W m^{-3}	500
ΔH_{fus}	Enthalpy of fusion for water	J kg^{-1}	33,430
ΔH_{vap}	Enthalpy of vaporization for water	J kg^{-1}	2,270,656
η	Porosity of the rock matrix	m^3 m^{-3}	0.04
T_0	Initial medium temperature	K	270
$t_{1/2}$	Effective half-life of radionuclides	s	109
r_c	Container radius	m	1
h_c	Height of the capsule in the case of cylinder	m	1
x_l	Half-length of the selected area by lateral	m	99
z_l	Height of the selected area	m	99
z_0	Position of the capsule center by height	m	50
t_{big}	Time of the model process finish	s	3.15 × 10^9
n_x	Node width	m	99
n_z	Node height	m	99
τ	Time step	s	500

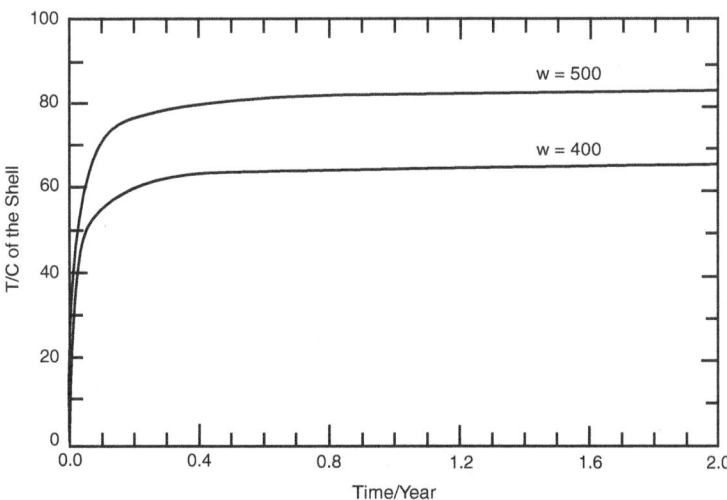

Figure 8.3 Predicted temperature at the surface of the shell.

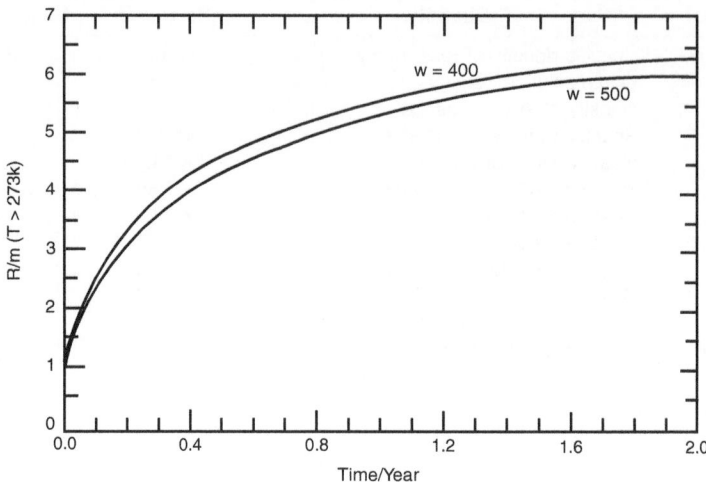

Figure 8.4 Predicted volume of unfrozen rock surrounding a radioactive waste container.

The spent nuclear fuel is to be disposed of in steel containers in trenches at a depth of about 20 m. For the first 1 to 250 years, the radioactivity is mainly due to the fission products. In approximately 500 years, contributions of fission products and actinides will be equal. In approximately 1000 years, actinides will be the main contributor to the radioactivity. The heat generation is within the range 0.2 to 2 $kW \cdot m^{-3}$. Uranium is present in the fuel as metallic uranium or as UAl_4 (plutonium as a metallic phase).

Kazakov et al. (1994) conducted numerical studies of heat conditions of isolated RAW in permafrost rocks. They investigated energy transfer with freezing and thawing as affected by the heat generation, shape of the heat source, and thermal conductivities of rocks. They showed that the radius of the melted zone could reach 20 m in about 40 to 60 years. Depending on the character of the radionuclides, this zone may persist for 100,000 years, slowly decreasing in its volume with a decrease in heat generation. The temperature on the surface of the container can reach 90°C. A new model is under development that includes convective energy and mass fluxes in the pore space of the thawed zone. Since convective fluxes can affect the shape of the thawed zone, a two-dimensional, rather than a one-dimensional, modeling approach is needed. Very preliminary estimates of the velocities of water fluxes along the container walls give values of about 1 to 1000 mm day^{-1} initially, decreasing to much smaller values at steady state.

Table 8.2   Average Chemical Composition of Novaya Zemlya Archipelago Limestone

Oxide	SiO_2	CaO	MgO	Fe_2O_3	Al_2O_3	SO_3	Lost at Heating (CO_2)
Content, wt%	5.88	43.65	4.50	0.95	3.89	0.99	39.65
Content, mol/kg	0.97	7.78	1.12	0.65	0.38	0.12	8.97

Approach for Modeling Chemical Interactions

For modeling chemical interactions that occur in various parts of the system, we use the original program *SLGSol* for calculating the equilibrium composition of multicomponent heterogeneous systems by the Gibbs energy minimization method. The program can also calculate sorption–desorption and ion-exchange equilibria. The input data for such calculations are temperature, pressure, bulk chemical composition of the system, and Gibbs energies of minerals and aqueous species. In such a manner, input data needed for calculation of chemical equilibrium in any time-space element are the results of solving the energy and mass tasks.

For future coupling of this program with the energy and mass transfer model, it will be necessary to calculate chemical equilibria extremely fast. To accomplish this, retaining the current reaction basis (independent chemical components) and their chemical potentials and using these as the first approximation for computing the chemical equilibrium at the next time step for any element (block of rock with the pore solution) is sensible. Accordingly, the input and the output routines of the program *SLGSol* were modified and additional linear transformations were added to the main program. This allows for the calculation of current chemical equilibrium composition within one to three iterations.

Thermodynamic Data

The thermodynamic DIANIK database was the main source of thermodynamic information on aqueous solution species and minerals. It contains data on more than 500 solids, including main rock-forming minerals, and uses the HKF model for calculating Gibbs energies of about 600 aqueous species over wide ranges of temperature and pressure (Khodakovsky et al., 1992).

Thermodynamics of Uranium and Plutonium

For calculating Gibbs energies of uranium and plutonium aqueous species and solids, the equilibrium constants for complexation, hydrolysis, oxide solubilities, and oxidization were collected and selected from the literature (Fuger and Oetting, 1976; Lemire and Tremaine, 1980; Devina et al., 1982; Devina et al., 1983; Kim, 1986; Cox et al., 1989; Sergeeva et al., 1989; Fuger et al., 1992; Fuger, 1992; Khodakovsky et al., 1992). For some reactions, one or more equilibrium reaction constants differ significantly. In some cases, the chosen data for aqueous complexes have not been evaluated critically. In such situations we tended to choose the constants that provide the maximum migration (i.e., yielded the worst-case estimate) of uranium and plutonium in the model. The data on uranium and plutonium used for computation of chemical equilibria are listed in Table 8.3.

Main Features of Chemical Interactions in the System

For modeling chemical equilibria in the system, we took a 13-component system consisting of H, O, Ca, Mg, Na, Fe, C, Cl, S, Al, Si, U, and Pu. Thirty-three solid

Table 8.3   Standard Gibbs Energies of Formation ($\Delta G_f°$, J mol⁻¹) of Some Uranium and Plutonium Solids and Aqueous Solution Species Used in the Model

| | Temperature, K | | | |
Solid	273.2	298.15	333.15	372.2
Uranium	1222	0	−1810	−3956
Plutonium	1366	0	−2029	−4449
UAl_4		−117121	−121893	−127861
UAl_3		−107017	−111940	−117823
UAl_2		−90227	−93883	−98297
UO (cr,cub.)		−395242	−396491	−398072
UO_2 (cr)	−1029980	−1031835	−1034655	−1038102
$UO_2(OH)_2$ (cr)	−1388542	−1391198	−1395409	−1400730
Schoepite	−1624921	−1628773	−1634776	−1642249
UO_3 (cr,mono)	−1143429	−1145741	−1149264	−1153574
U_2O_3 (cr,hex)	−1121039	−1123291	−1126798	−1131190
U_3O_7 (cr,tet)	−3234698	−3240721	−3249910	−3261170
U_3O_8 (cr,rho)	−3359039	−3365502	−3375385	−3387508
U_4O_9 (cr,cub)	−4268531	−4276532	−4288801	−4304030
PuO (cr)	−536849	−538556	−541132	−544239
PuO_2 (cr)	−996030	−997607	−1000056	−1003100
$Pu(OH)_2CO_3$		−1610202		
Pu_2O_3 (cr,cub.)	−1632246	−1635930	−1641565	−1648459
Species				
$U(OH)^{3+}$ (ao)		−1236000	−1217011	−1216793
$U(OH)_4$ (ao)		−1448600	−1431262	−1438354
UO_2^{+2} (ao)		−952950	−949043	−944874
UO_2OH^+ (ao)		−1157100	−1158162	−1159933
$(UO_2)_2(OH)_2^{+2}$		−2348800	−2349053	−2350448
$UO_2(OH)_2$		−1357000	−1353982	−1355980
UO_2CO_3 (ao)		−1537300	−1539079	−1541899
$UO_2(CO_3)_2^{-2}$		−2104700	−2110258	−2117908
$UO_2(CO_3)_3^{-4}$		−2658940	−2658563	−2657103
Pu^{+3} (ao)		−599218	−591903	−584439
Pu^{+4} (ao)		−481600	−469480	−452799
$Pu(OH)^{+3}$		−709118	−709253	−702941
$Pu(OH)_2^{+2}$		−935493	−942647	−938793
$Pu(OH)^{3+}$		−1153877	−1163282	−1167500
$PuCO_3^{+2}$		−1079799	−1212091	−1217860
$Pu(CO_3)^2$		−1671147		
$Pu(CO_3)_3^{-2}$		−2236805		
$Pu(CO_3)_4^{-4}$		−2780770		
PuO^{2+} (ao)		−850477	−850087	−848051
$PuO_2(OH)$		−1031647	−1038826	−1041032
PuO_2^{+2} (ao)		−756900	−754338	−750010
$PuO_2(OH)^+$		−961354	−963490	−964503
$(PuO_2)_2(OH)_2^{+2}$		−1942126	−1941653	−1936132
$PuO_2(OH)^{3-}$		−1325276		
$PuO_2(CO_3)_2^{-2}$		−1899189	−1907837	−1908276

phases were considered: calcite, dolomite, siderite, magnesite, ferrodolomite, kaolinite, quartz, daphnite, clinochlore, laumontite, gibbsite, goethite, hematite, magnetite, FeO (cr, cubic), iron, aluminum, pyrite, gypsum, uranium, UO, UO_2, $UO_2(OH)_2$, shoepite, UO_3, U_3O_7, U_3O_8, U_4O_9, reserfordite, plutonium, PuO_2, and Pu_2O_3. In addition, 60 aqueous species were chosen.

Chemical Compositions of the Pore Solutions

The chemical composition of the pore aqueous solutions in equilibrium with the limestone from 0 to 99°C has been calculated. The mass balance for calculations consisted of 1 kg of the limestone of the specified chemical composition (Table 8.2) plus a mass of water that occupied 4% of the limestone's volume. The system was supposed to be open with respect to CO_2. As a first approximation, partial pressure of CO_2 was taken one order of magnitude higher than the atmospheric carbon dioxide pressure. The calculated equilibrium mineral assemblages consisted of calcite, dolomite, kaolinite, chlorite, and a small amount of pyrite, and are in good accordance with mineralogical data.

These pore solutions have the highest concentrations of calcium, magnesium, and ferrous species (Table 8.4), as well as the lowest oxidizing potential. It should be emphasized that concentrations of calcium, magnesium, and ferrous iron decrease

Table 8.4   Some Parameters of Chemical Composition of the Pore Aqueous Solution of the Limestone According to Equilibrium Calculations

Content, M	Temperature, K		
	273.15	293.15	372.15
$CO_2°$(aq)	2.9×10^{-4}	1.1×10^{-4}	3.4×10^{-5}
ΣCa	1.36×10^{-3}	9.8×10^{-4}	4.1×10^{-5}
ΣMg	6.2×10^{-4}	3.01×10^{-4}	1.75×10^{-6}
ΣFe	1.4×10^{-6}	1.7×10^{-7}	5.1×10^{-9}
pH	7.70	7.73	8.13

with heating. This means that a precipitation of carbonates (and decreasing of porosity of the limestone) should take place near the heat-generating container.

It should be noted that the content of chloride, sodium, and potassium in the rock had not been determined. Nevertheless, a knowledge of these parameters may be important from the viewpoint of the unfrozen brine formation during future low fractional freezing of dilute pore solution.

Corrosion of the Steel Container by Pore Solutions

According to the electrochemical theory of corrosion at anodic sites of an iron surface, the following reaction takes place:

$$Fe = Fe^{2+} + 2e^-$$ (Reaction 1)

The rate of this reaction is found to be dependent upon the rate of the cathode reaction. Either of the two reactions is typical of the cathodic sites:

$$2H^+ + 2e^- = H_2 \qquad\qquad \text{(Reaction 2)}$$

$$2H^+ + (1/2)O_2 + 2e^- = H_2O \qquad\qquad \text{(Reaction 3)}$$

Reaction 2 is fairly rapid in acids, but very slow in alkaline or neutral media, like the pore solutions of the limestone. It can be accelerated by the presence of dissolved oxygen as depicted in Reaction 3. The rate of this reaction is proportional to the diffusion of oxygen to the metal surface, which is, in turn, proportional to the concentration of dissolved oxygen in the aqueous solution. According to equilibrium calculations of the chemical composition of the pore solutions of the limestone, the concentration of dissolved oxygen is extremely low due to the presence of pyrite. It is less than $5 \cdot 10^{-16}$ or about 2×10^{-14}‰ O_2 (without taking into account diffusion or another flow of atmospheric oxygen to the repository). Thus, the chemical reaction of steel dissolution by aqueous solutions, with very low contents of dissolved oxygen and low concentrations of the H^+ ion, takes place at a very slow rate and by the following mechanism at given conditions:

$$Fe + 2H_2O = Fe^{+2} + 2OH^- + H_2 \qquad\qquad \text{(Reaction 4)}$$

According to the data from Whitman et al. (1924) cited by Uhlig (1948), the corrosion penetration in steel does not exceed 9.1 µm year^{-1} per ‰ O_2 at 40°C and 5 µm year^{-1} per ‰ O_2 at 22°C within the pH range 4.5 to10. If we recalculate these rates to the calculated oxygen content and extrapolate these to 90°C, it is possible to estimate, very roughly, the corrosion rate for steel in this environment. It shouldn't exceed about 2.5 nm year^{-1}.

The equilibrium calculations for the given CO_3^{2-} rich conditions suggest that the following reaction should govern the process of the iron container corrosion:

$$4Fe + 7H_2O + 5CO_3^{2-} = 4FeCO_3 + CH_4 + 10OH^- \qquad \text{(Reaction 5)}$$

Apparently the chemical reaction of methane formation has severe kinetic restrictions. Because of this, dissolved CH_4 was deleted from the system as a chemical component. After this procedure, we determined that the following chemical reaction should be the dominant Fe dissolution mechanism:

$$Fe + 2H_2O + CO_3^{2-} = FeCO_3 + H_2 + 2OH^- \qquad \text{(Reaction 6)}$$

Iron oxidation results in the deposition of siderite. The formation of the patina of a mineral as dense as siderite on the container's surface would isolate the interior from contact with the pore solutions. According to our calculations of multicomponent equilibria, this phenomenon would be followed by the precipitation of calcium and magnesium carbonates from elevated temperatures due to the heat generated by the container's contents. The precipitation of carbonates near the container surface should reduce the porosity, and therefore permeability, of the limestone rock adjacent to the container. From these points of view the choice of limestone as a wall rock seems to be successful.

Uranium and Plutonium Dissolution

Because of the presence of metallic Fe (container) and pyrite in the rock, the oxidation potential would be very low. Accordingly, the solubility of uranium solids is very low. The uranium concentration in the pore solution is about 1×10^{-12} mol kg^{-1}(U[OH]$_4^0$ is its dominant species). Plutonium concentrations are much higher, about 1×10^{-4} mol kg^{-1}; its dominant species are Pu(IV)CO$_3^{2+}$ and Pu^{+3}. Taking into account a small amount of aqueous solution, most of the uranium and plutonium masses are in solid phases. As an example, in Table 8.5, one can see the result of calculation of the equilibrium composition of the system, consisting of the limestone with the pore solution, iron, and UAl$_3$ and metallic plutonium at 99°C and at the partial pressure of H$_2$ equal to 10 mbar. One can see that the radionuclides were transformed to oxides.

Hydrogen gas is produced by the oxidation of metals (e.g., Fe, U, and Pu) by water. Dissolved hydrogen gas in pore solutions would tend to retard corrosion of a steel container. Hydrogen production and diffusion in water-saturated rock would be a complex dynamic system, warranting more detailed study.

Future Development of the Model

The following improvements in the model are planned:

- To include the chemical behavior of strontium and cesium, which contribute the majority of the radioactivity during the first hundred years of isolation
- To include radionuclide-sorption reactions on rock-forming minerals and their co-precipitation with calcium, magnesium, and ferrous iron
- To estimate the scale of diffusion transport of the radionuclides in the permafrost rock pore space
- To include the short half-life radionuclides ^{90}Sr and ^{137}Cs in the model
- To estimate the effect of hydrogen and oxygen diffusion on the rate of steel container corrosion
- To combine the chemical interaction module with the heat-mass transfer module

ACKNOWLEDGMENTS

Financial support is acknowledged from the Russian Foundation of Basic Research (Project 97-05-64197); the U.S. Army Cold Regions Research and Engineering Laboratory (Work Unit 61102/AT24/129/EE005, entitled "Chemistry of Frozen Ground"); the U.S. Army Environmental Quality Basic Research Enhancement Program (Work Unit AT25-EC-F02, entitled "Thermodynamic Data for Aque-

Table 8.5   An Example of the Calculation of the Equilibrium Composition of the System Spent Nuclear Fuel (UAl3 + Pu)–Steel Container (Fe)–Limestone–Pore Water at 99°C with the Program *SLGSol*

TK = 372.20 pressure 1.00000 bar
It was taken into account: solid phases — 30
species — 57
The system is open with respect to 1 component:

comp.	lg(act)	activity
H_2 (ao	-4.00	.10E-03

Solution
ion str. = .04000

n		Species	Mole	Molality	Activity	Act. Coeff.	Free Energy (J/mol)
1	a:	H_2	.12758E-05	.99578E-04	.10000E-03	.0042	12159
2	a:	OH^-	.46417E-08	.36229E-06	.29305E-06	.8089	-155545
3	a:	Cl^-	.20000E-04	.15610E-02	.12627E-02	.8089	-134533
4	a:	HS^-	.27496E-12	.21461E-10	.17359E-10	.8089	7614
5	a:	H_2S	.12367E-11	.96529E-10	.96938E-10	1.0042	-38613
6	a:	CO_2	.43075E-03	.33620E-01	.33763E-01	1.0042	-396504
7	a:	CO_3^{2-}	.10422E-07	.81341E-06	.34382E-06	.4227	-522125
8	a:	HCO_3^-	.12097E-03	.94417E-0	.76371E-02	.8089	-594002
9	a:	H_2CO_3	.51120E-03	.39900E-01	.40069E-01	1.0042	-640036
10	a:	SiO_2	.10268E-03	.80145E-03	.80485E-03	1.0042	-837708
11	a:	Al^{3+}	.22838E-15	.17825E-13	.30853E-14	.1731	-460355
12	a:	$AlOH^{2+}$	.35721E-13	.27881E-11	.12316E-11	.4417	-680988
13	a:	$Al(OH)_2$	.44857E-11	.35011E-09	.28320E-09	.8089	-899914
14	a:	$Al(OH)_3$	.55827E-10	.43573E-08	.43758E-08	1.0042	-1110484
15	a:	Fe^{2+}	.95520E-07	.74555E-05	.32934E-05	.4417	-83388
16	a:	$FeOH^+$	.76942E-09	.60054E-07	.48576E-07	.8089	-272437
17	a:	$FeOH^{2+}$	.14640E-18	.11427E-16	.50477E-17	.4417	-234031
18	a:	$U(OH)_3$	.27173E-16	.21209E-14	.17155E-14	.8089	-1216793
19	a:	$U(OH)_4$	.11792E-13	.92039E-12	.92429E-12	1.0042	-1438354

n		Species				Vol %	
20	a:	UO_2CO_3	.76365E-20	.59604E-18	.59856E-18	1.0042	-1541899
21	a:	$UO_2(CO_3)_3^{2-}$	.22748E-18	.17755E-16	.75049E-17	.4227	-2117908
22	a:	$UO_2(CO_3)_3^{-4}$	.67230E-21	.52474E-19	.64153E-21	.0122	-2657103
23	a:	Pu^{+3}	.89560E-10	.69902E-08	.12099E-08	.1731	-584439
24	a:	$Pu(OH)_3$	.26828E-18	.20939E-16	.16937E-16	.8089	-1167500
25	a:	$PuCO_3^{+2}$	.12790E-05	.99830E-04	.44100E-04	.4417	-1217860
26	a:	Mg^{2+}	.58346E-05	.45540E-03	.20117E-03	.4417	-443663
27	a:	Ca^{2+}	.53271E-04	.41578E-02	.18367E-02	.4417	-548351
28	a:	$CaOH^+$	.63423E-09	.49503E-07	.40042E-07	.8089	-717232
29	a:	Na^+	.20000E-04	.15610E-02	.12627E-02	.8089	-266597
30	l:	$H_2O(liq)$	.71002	.99835E+00	.99835E+00	1.0000	-242997
31	a:	H^+	.28798E-07	.22477E-05	.18181E-05	.8089	0

pH = 5.7404

Solid Phases

n		Phase	Mole	G (J/mol)	Vol %
1	c:	$PuO_2(cr)$	.49987209E-02	-1003100.	.032
2	c:	$UO_2(cr)$	.10000000E-01	-1038102.	.067
3	c:	kaolinite	.19639950	-7633214.	10.697
4	c:	dolomite	1.1199942	-2174216.	19.719
5	c:	daphnite	.22010000E-02	-6580604.	.129
6	c:	calcite	6.6599526	-1136030.	67.304
7	c:	pyrite	.60000000E-01	-244234.	.393
8	c:	siderite	.68994904E-01	-690637.	.555
9	c:	quartz	.17778873	-859757.	1.104

volume 365.43560 cm³

Balance

	Common	Solids	Fluid	Common Comp.
Fe	.140000E+00	.140000E+00	.962898E-07	.140000E+00
U	.100000E-01	.100000E-01	.118195E-13	.100000E-01
Pu	.500000E-02	.499872E-02	.127912E-05	.500000E-02

ous Solutions at Low Temperatures"); and the Radioactive Waste Management Program, Office of International Affairs, National Research Council.

REFERENCES

Bird, R.B., W.E. Stewart, and E.N. Lightfoot. *Transport Phenomena*, Wiley, New York, 1960.

Carslaw, H.S. *Introduction to the Mathematical Theory of the Conduction of Heat in Solids*, 2nd. ed., Dover, New York, 1945.

Cox, G.D., D.D. Wagman, and V.A. Medvedev. *CODATA Key Values for Thermodynamics*, Hemisphere Publishing, New York, 1989.

Devina, O.A., M.Ye. Yefimov, V.A. Medvedev, and I.L. Khodakovsky. Thermodynamic properties of the uranyk ion in aqueous solution at elevated temperatures, *Geochem. Int.*, 19(5), 161–172, 1982.

Devina, O.A., M.Ye. Yefimov, V.A. Medvedev, and I.L. Khodakovsky. Thermochemical determination of the stability constant of $UO_2(CO_3)_3^{4-}$ (sol) at 25–300°C, *Geochem. Int.*, 20(3), 10–18, 1983.

Fuger, J. Thermodynamic properties of actinide aqueous species relevant to geochemical problems, *Radiochim. Acta*, 58/59, 81–91, 1992.

Fuger, J. and F.L. Oetting. *The Actinide Aqueous Ions, II. The Chemical Thermodynamics of Actinide Elements and Compounds*, International Atomic Energy Agency, Vienna, 1976.

Fuger, J., I.L. Khodakovsky, E.I. Sergeyeva, V.A. Medvedev, and J.D. Navratil. *The Actinide Aqueous Inorganic Complexes, XII, The Chemical Thermodynamics of Actinide Elements and Compounds*, International Atomic Energy Agency, Vienna, 1992.

Harris, J.S. *Ground Freezing in Practice*, Thomas Telford, London, 1995.

Kazakov, A.N. Underground isolation of radioactive wastes in permafrost rocks, *Geoekolgiya: Izhenernaya Geologiya. Gidrogeologiya. Goekriologiya*, 6, 71–73, 1996 (In Russian).

Kazakov, A.N., N. G. Schetinin, and N.F. Lobanov. Spherical-symmetrical description of thermal-physical processes in the system: buried radioactive waste-perennially frozen bedrock mass, in A.N. Kazakov and V.V. Lopatin, Eds., *Underground Isolation of Radioactive Waste in Perennially Frozen Bedrock Mass*, Central Institute for Management, Economics and Information, Ministry of Atomic Energy, Russian Federation, Moscow, 1994 (in Russian).

Khodakovsky, I.L. DiaNIK-Thermodynamic databases of minerals and calculations of equilibrium compositions of natural systems, in *Abstracts of the 12th IUPAC Conference on Chemical Thermodynamics*, Snowbird, Utah, 1992, 157–158.

Khodakovsky, I.L., A.G Volosov, Yu.V. Semenov, et al. Development of the database and a sofware for computer modeling of the processes of regional geomigration, Technical Report, Vernadsky Institute of the Russian Academy of Sciences, Moscow, 1992, 170 p. (in Russian).

Kim, J.I. Chemical behavior of transuranic elements in natural aquatic systems, in A.J. Freeman and C. Keller, Eds., *Handbook on the Physics and Chemistry of the Actinides*, Vol. 4, North-Holland, New York, 1986, 413–455.

Lemire, R.J. and P.K. Tremaine. Uranium and plutonium equilibria in aqueous solutions to 200°C, *J. Chem. Eng. Data*, 25, 361–370, 1980.

Sergeeva, E.I., O.A. Devina, K.S. Gavrichev, V.E. Gorbunov, V.M. Gurevich, M.E. Efimov, V.A. Medvedev, I.L. Khodakovsky. Investigation of the UO_3-CO_2-H_2O system with the use of calorimetric, spectrophotometric, and solubility methods, in B.N. Laskorin and B.F. Miysoedov, Eds., *Chemistry of the Uranium*, Nauka, Moscow, 1989, 36–41 (in Russian).

Shapkin, A.I. A variational method of calculating convection in a magma chamber, *Geochem. Int.*, 29(2), 134, 1992.

Shapkin, A. I. A model for the flow of the settling phase in a magmatic chamber, *Geochem. Int.*, 30(7), 41–52, 1993.

Sumgin, M.I., C.P. Kachurin, N.I. Tolstikhin, and V.F. Tumel. *General Permafrostology*, Izd-vo Akademii nauk SSSR, Moscow, 1940 (in Russian).

Tikhonov, A.N. and A.A. Samarskii, *Equations of Mathematical Physics* (translated by A.R.M. Robson and P. Basu), Dover, New York, 1990.

Uhlig, H.H. Iron and steel, in H.H. Uhlig, Ed., *The Corrosion Handbook*, Wiley, New York, 1948, 125–143.

Whitman, W.G., R.P. Russell, and V.J. Altieri. Effect of hydrogen-ion concentration on the submerged corrosion of steel, *Ind. Eng. Chem.*, 16, 665–670, 1924.

CHAPTER 9

Permafrost and Stratigraphic Layer Identification Using a Hierarchical Neural Network for Interpretation of Ground-Penetrating Radar

John M. Sullivan, Jr., Reinhold Ludwig, and Dmitry V. Repin

CONTENTS

INTRODUCTION

Ground-penetrating radar (GPR) is an electromagnetic (EM) remote sensing technique that uses radio waves, typically in the 10- to 2500-MHz frequency range, to locate and map features and structures below the ground surface (BGS). In general, a GPR system transmits a short electromagnetic pulse into the ground. The pulse is reflected, refracted, or scattered by the targets that exhibit some difference in electrical properties (dielectric permittivity, conductivity, and magnetic permeability) and is then recorded by the receiving antennas. The greater the difference in the dielectric permeability, the larger is the amplitude of the reflection pulse.

High radar frequencies are needed to achieve a good spatial resolution, but penetration depth of the electric field is inversely proportional to the frequency. Hence the choice of frequency range is a trade-off between resolution and penetration depth. Penetration depth also depends on the nature of the soil, which has different attenuation properties. For example, desert sand has an attenuation of about 1 dB/m for a 1-GHz frequency, and clay has an attenuation of 100 dB/m at the same frequency.

The reflected wave is sampled with 512 or 1024 points taken through the region of interest. A typical recorded signal in the time domain, an A-scan, is shown in Figure 9.1. These A-scans are recorded consecutively along a spatial direction or transect line. A typical GPR system records 5 to 10 scans per meter. A variety of software packages exists for visualization and data processing. Frequently GPR soundings are performed by dragging a GPR hardware package, including transmitting and receiving antennas, behind a vehicle.

GPR has been used for various scientific, industrial, environmental, and military applications:

- Stratigraphic layers profiling (water table detection, etc.)
- Ice thickness measurements
- Buried object detection
- Archaeological investigations
- Road investigations
- Fracture detection in hard rock
- Liquid contaminant detection

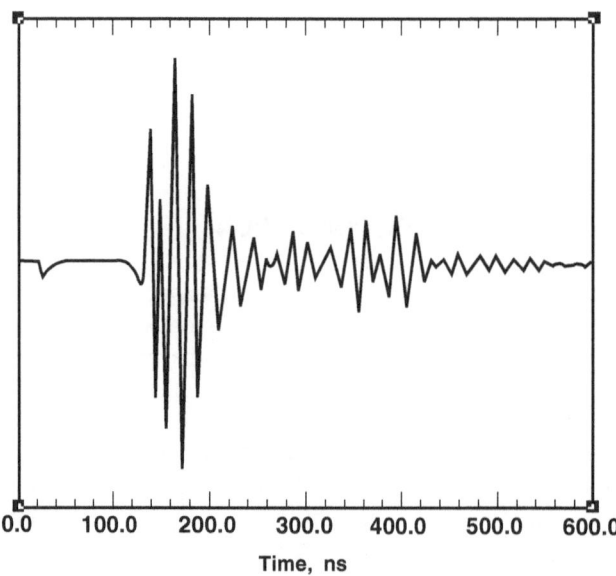

Figure 9.1 Typical GPR A-scan.

- Safety inspections at nuclear power plants
- Antipersonnel mine detection

This work concentrates on a particular application: noninvasive site characterization of stratigraphic layer depths in the vicinity of Fairbanks, Alaska. One of the most important features of this region is the presence of the permanently frozen materials (permafrost), which have distinct dielectric properties. The GPR data were collected to identify horizontal and vertical distributions of permafrost zones, the water table, and bedrock. GPR was chosen to accomplish this task, since the scales of depths and lateral variations of permafrost are too small to be resolved with seismic methods, and too large for efficient mapping with electromagnetic inductance methods (Arcone et al., 1998). Frozen soils (mostly sands and gravels) are low-loss propagation media for electromagnetic waves at radio frequencies. Therefore, saturated sediments form a continuous and highly reflective surface with frozen ground. In this situation GPR is one of the best tools to study the vertical distribution of the different subsurface materials.

The ability to characterize the subsurface media is important for environmental problems related to the fate and transport of contaminants within groundwater. This site characterization is critical where discontinuous permafrost exists. Subsurface channels and conduits are created along thaw zones with contaminant flow paths that are not intuitive based on traditional soil stratification identification. Successful GPR profiling plays a vital role in the detection of these permafrost zones.

Significant research efforts in the field of GPR were made by CRREL (U.S. Army Cold Regions Research and Engineering Laboratory). CRREL performed multibandwidth reflection profiling of discontinuous permafrost via GPR during 1993–1994. The GPR antennas' bandwidths were centered near 50, 100, and 300 MHz. An area spanning 8 km^2 had over 100 km of GPR profiles recorded (Arcone et al., 1998). The work presented herein used the 100-MHz center frequency GPR data with a 600-ns A-scan recording.

The subsurface layer configurations for the area are shown schematically in Figure 9.2. Variations in the permafrost structure may have natural (riverine) or

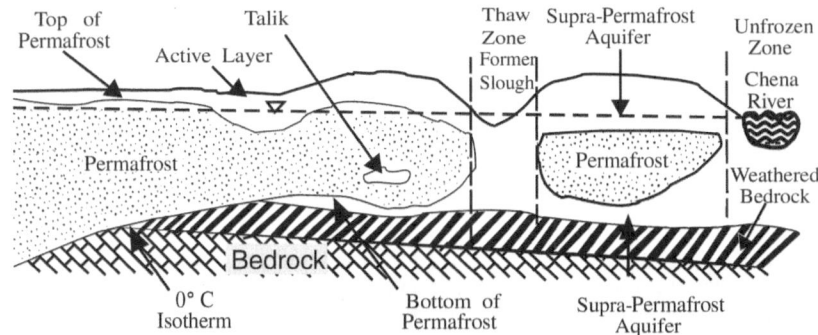

Figure 9.2 Possible configuration of subsurface layers. (From Arcone et al., 1998. With permission.)

artificial (roads or other human activities) origins, and those variations may be significant over a relatively short lateral scale. On the other hand, variations of the water table and the bedrock absolute depth are relatively small. Along with the above-mentioned major subsurface components, variations in types of soils and moisture content introduce additional difficulties into appropriate data processing.

The properties that characterize the propagation of the electromagnetic waves in an isotropic medium are the dielectric permittivity (ε) and the electric conductivity (σ). The first one is responsible for the wave propagation time, and the second one accounts for the loss factor. Both values are very sensitive to the soil moisture content (Smithsonian Institution, 1954; Parkhomenko, 1967; Arcone et al., 1998). For example, ε for dry soil can be in a range of 4 to 8, but with a moisture content of 30%, ε can be as high as 40. The soil conductivity affects mostly the EM pulse evanescence and has a small effect on the reflections from the layer boundaries. Dielectric permittivity values for soil types of our concern (dry, saturated, frozen, unfrozen) are presented in the Table 9.1.

Table 9.1 Soil Dielectric Permittivity

Soil Type	Permittivity Value
Permanently frozen materials	4.4–5.6
Unfrozen saturated sediments	12.0–45.0
Dry soil	4.0–8.0
Weathered bedrock	>11.0
Bedrock: granite, sandstone	7.0–9.0, 10.0
Sand: dry, 15% moisture, 25% moisture content	≈3.0, ≈9.0, ≈25.0

Previously GPR line data processing used RADAN™ software to visualize sets of adjacent A-scans, coupled with a human-expert evaluation based on previous knowledge and generalization from the available geophysical data. This analysis was augmented with the information of the subsurface structure from a number of boreholes (Lawson et al., 1996). Those boreholes were drilled throughout the area, and types and properties of materials, as well as corresponding layer depths, were recorded.

This traditional method is highly subjective. It lacks significant automation, and the layer interfaces identified by the expert still lack depth information since the record is a time history. The time scale of the GPR signal does not scale linearly with depth. Consequently, we have targeted a novel strategy for automatic stratigraphic layer identification and depth predictions.

The novel strategy has the following features:

- The main task is split into several consecutive stages, decreasing the degree of uncertainty within each step.
- Data adaptive techniques are applied to the GPR signal, transforming it into a highly informative and easily interpretable feature vector.
- Neural network modules used provide high error and noise tolerance.

A block diagram of the approach is shown in Figure 9.3. The major processing units of the data processing system are neural networks (NN). One NN (NN1) is

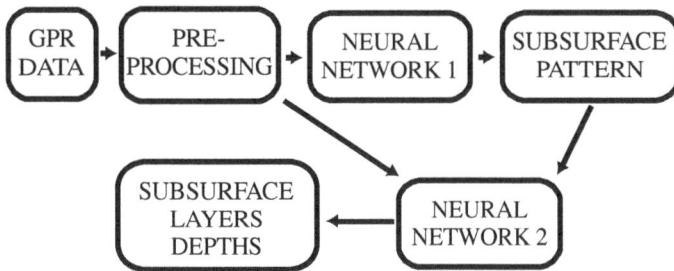

Figure 9.3 Block diagrams of the data processing architecture.

designed to perform best on a classification problem. The second NN (NN2) provides good approximation ability for depth analysis within the classified problem. Both neural networks belong to the class of two-layer feedforward networks. Both neural networks are of the multilayer perceptron (MLP) type trained with back-propagation or scaled conjugate gradient algorithms (Haykin, 1994).

Another important part of this architecture is the preprocessing stage — it has four separate blocks responsible for proper data handling and feature extraction to provide the neural networks with relevant input information.

In general, the system operation may be described as follows. The preprocessing unit performs checking of the data consistency; the second stage decomposes the initial A-scan with our adaptive transform (AT) technique into a set of data adaptive basis functions; the third section incorporates information about the previous A-scans with the linear regression method into the current processing step. Then a feature vector forms from the coefficients of the AT decomposition, and some other prior available information is fed into the neural network 1 (NN1) to recognize the "subsurface pattern" for the current A-scan. This subsurface pattern reflects one of the possible subsurface layer configurations. During the next step, part of the transformed input for the NN1 and already available information about the subsurface pattern are used as input for neural network 2 (NN2), which produces the set of the stratigraphic layer depths.

A feedback routine is used to account for possible incorrect recognition of the pattern or other "alarm" signals produced during system operation. This feedback feature adds flexibility to the entire setup that already has high noise/error-tolerant processing units as neural networks.

Our main preprocessing goal is to reduce the dimension of the problem. A typical A-scan is represented with 512 to 1024 sampling points; this A-scan is too large to be used for input into a neural network. The strategy is to extract the most discriminative characteristics of the data, while preserving the chronological order of reflection pulses contained in the A-scan. To accomplish this we have developed an adaptive transform routine that decomposes the actual A-scan, $F(t)$, into a set of initial pulses:

$$G(t) = \sum_{i=1}^{K} w_i I_i \left(\frac{t - C_i}{S} \right) \tag{9.1}$$

where $G(t)$ is the reconstructed signal from the AT coefficients, K is the total number of functions in the decomposition, w_i are the weight values, $I_i(t)$ the initial pulse (IP), c_i are the shifts, and s is the dilation value to account for dispersion effects (Repin, 1998).

A low-dimensional feature vector was constructed from the AT coefficients and used as an input for the neural network processing units. Because of the different formulations (classification and approximation) of NN1 and NN2, different features of the set of AT coefficients were chosen for data representation. For the classification-oriented neural network, the first ten largest-in-magnitude weights w_i in chronological order were chosen. Then the same weights were arranged in descending order of their magnitude and appended to the chronological set. Shift values were not used directly; only the order of the particular weight appearance had importance. Consequently, NN1 had 20 inputs instead of the original 512 or 1024.

The implications for such a choice of the feature vector are the following: (1) the layers of the single subsurface pattern may be located at different depths; (2) the particular time of the reflection is not useful, but the chronological order of the weights and their values that contain the information about the sign of the reflection (whether it is inverted or not) is useful; and (3) its strength can help characterize some subsurface layer configurations. The same weights in the order of magnitude provide some additional information that may become necessary when, for example, a wrong, but not very large, coefficient is identified by the AT. In this case the chronological order of weights is disturbed, but the magnitude order remains the same allowing the system to perform correct pattern recognition.

The approximation-oriented NN2 does not require the weight values for the input, since the subsurface pattern is already identified and the number and order of layers are known. Now, the exact shift values contain the essential information about the location of the particular reflections, and ten of them that correspond to the first ten weights exceeding a certain threshold value are used to construct the feature vector. The construction of the feature vector from adaptive transform coefficients is not unique. However, the training sets generated using the above techniques applied to NN1 and NN2 demonstrated very good performance.

Each neural network was based on the MLP model, which is trained with supervised algorithms. It belongs to the class of feedforward layered neural networks — it has an input layer, output layer, and one or several internal or hidden layers. Connections in this type of network exist only between the neurons of different layers, and each neuron is connected to all the neurons in the previous and following layers. There are no lateral (within one layer) connections or feedback. The input signal propagates through the network in a forward direction on a layer-by-layer basis as in Figure 9.4.

MLPs found their application in different technological and other areas. Each processing unit (neuron) of the MLP network can be described with an activation or transfer function. The essential feature of that function is nonlinearity and for many applications it may be similar to Equations 9.2 or 9.3:

$$f(u_i) = \frac{c_1}{1 + \exp\{-u_i\}} \tag{9.2}$$

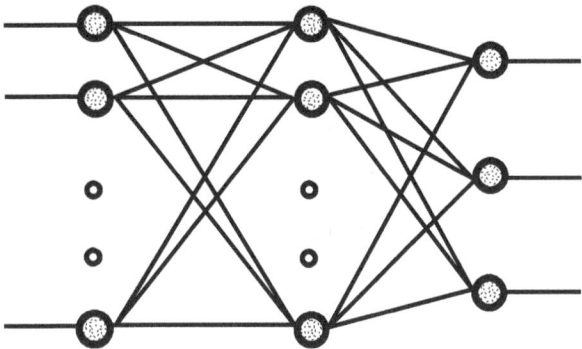

Figure 9.4 MLP network.

$$f(u_i) = c_1 \frac{u_i}{|u_i| + c_2} \tag{9.3}$$

where c_1, c_2 are constants and u_i is the output of the neuron before applying the nonlinearity:

$$u_i = \sum_{j=1}^{M} w_{ij} \cdot x_j + \theta_i \tag{9.4}$$

where x_j is the network input or the output of the previous layer, w and θ are the weights and the bias corresponding to this neuron, respectively. Input propagation through the network is a forward run; finally it produces a set of output-activation signals of the output layer units. During the forward run all the weight and biases values remain fixed. During the backward run, weights and biases are adjusted according to the error-correction learning rule. The error is defined as a sum of squared differences of the activation values of output neurons and the target output values provided by the training set. The error propagates backward through the network, and the weights and biases are adjusted to reduce the cumulative error. This algorithm is a "Back Propagation of Error" learning algorithm.

RESULTS

For better understanding of the underlying physics associated with the electromagnetic pulse propagation through the ground, a finite difference time domain (FDTD) simulator was developed (Plunkett, 1997). It used a plane wave GPR excitation with good agreement to actual GPR recordings, as shown in Figure 9.5. This numerical formulation was used to develop a complete training set for the neural network.

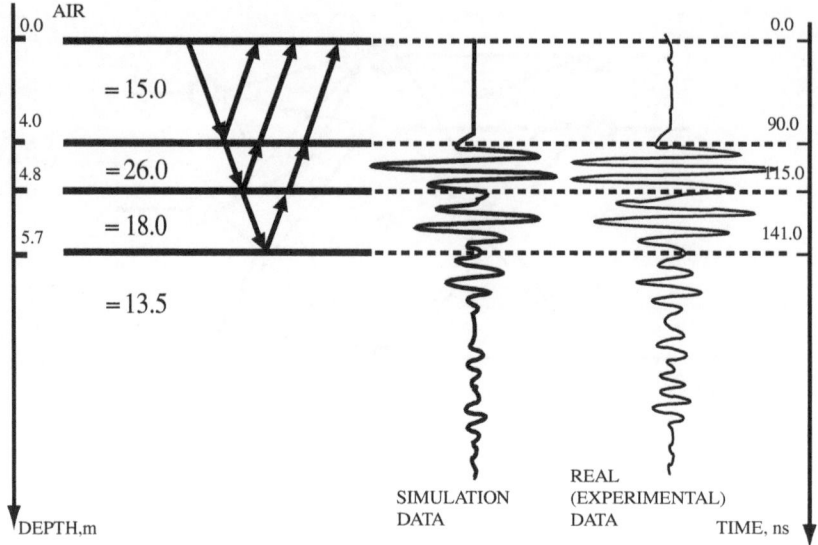

Figure 9.5 Depth-time diagram for simulated and real scans.

The synthetic training set (main training set) consisted of approximately 3300 artificial scans representing seven possible subsurface patterns. Depths of the layers were varied to span the depth variations observed in the field. Specific subsets corresponding to patterns 1 through 7 were used for training each of the seven NN2 models, respectively.

Synthetic GPR Line Based on CR93-11

The first 600 m of a transect line (CR93-11) (Arcone et al., 1998) traveling west to east was approximated with the synthetic geometry shown in Figure 9.6. The subsurface exhibited four different patterns along this transect line. Patterns 1 and 2 look the same in the figure, but the depth to bedrock in pattern 1 is beyond GPR detection.

NN1 was trained with the main training set. Once trained it identified each subsurface pattern correctly. The depth predictions for patterns 1, 2, and 3 (using corresponding NN2 modules) are shown in Figures 9.7 to 9.9, respectively. The system had some difficulty identifying the permafrost depth in pattern 1 for the initial scans. We are currently examining this questionable region. However, the system did an admirable job predicting all other scans and patterns as delineated in Figures 9.7 to 9.9.

Synthetic GPR Line Based on CR94-61r

A second transect line traveling south to north (CR94-61r) was simulated for the first 900 m. The same, previously trained, NN1 identified each subsurface pattern correctly for the geometry displayed in Figure 9.10. Depth predictions for patterns

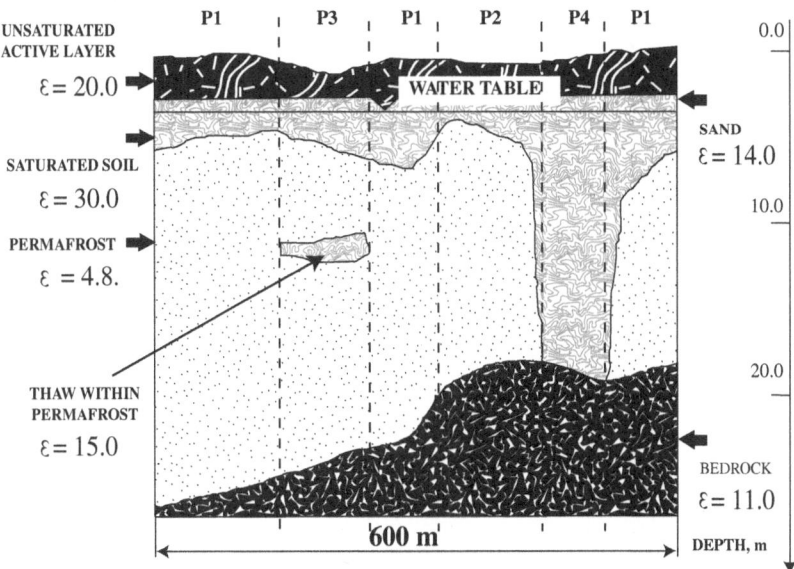

Figure 9.6 Geometry and dielectric constants for synthetic CR93-11 GPR line.

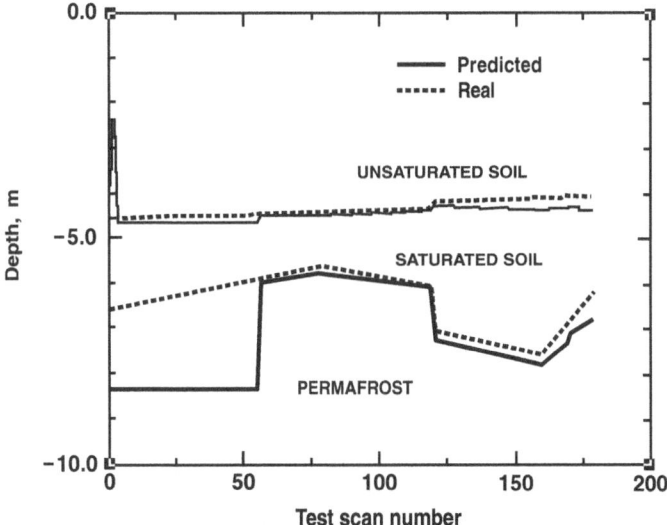

Figure 9.7 Depth predictions for pattern 1.

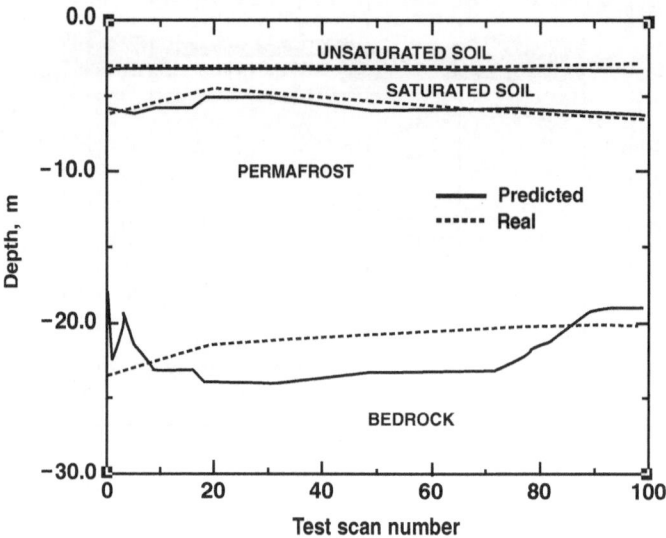

Figure 9.8 Depth predictions for pattern 2.

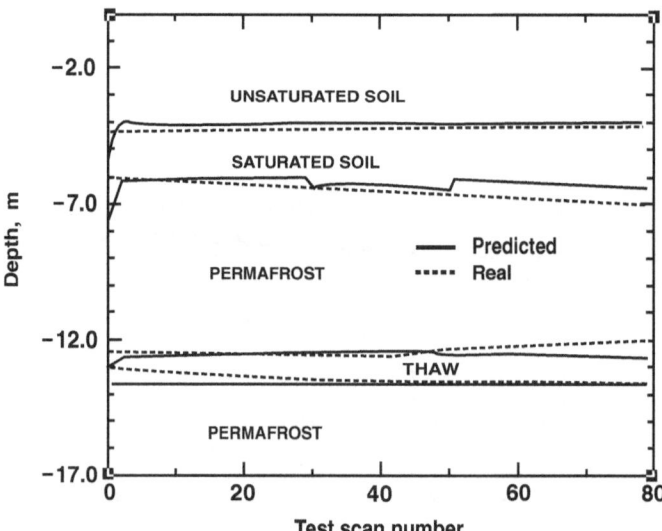

Figure 9.9 Depth predictions for pattern 3.

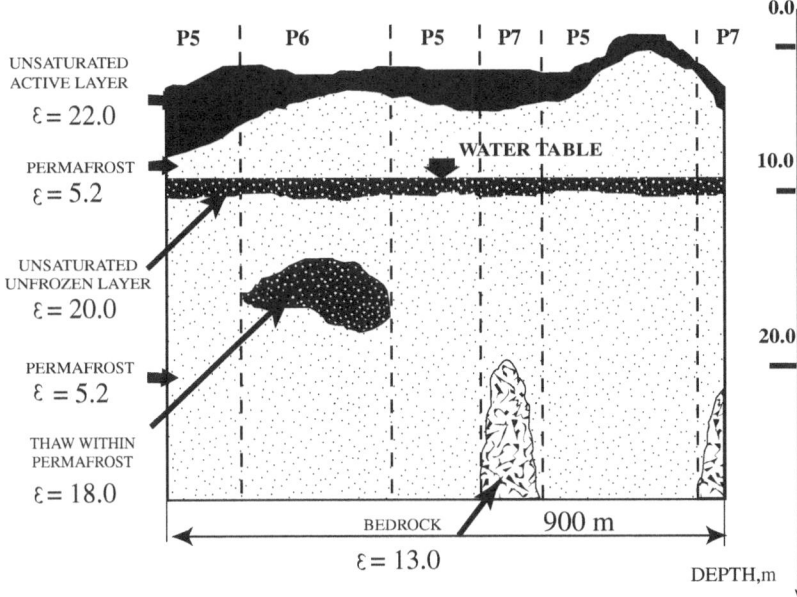

Figure 9.10 Geometry and dielectric constants for synthetic CR93-61r GPR line.

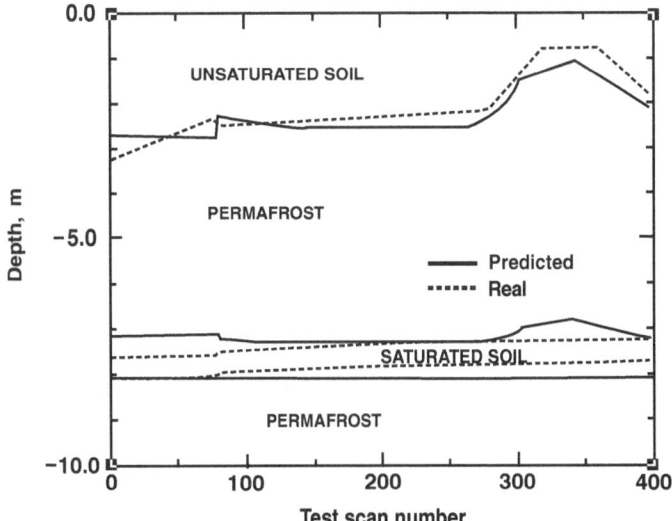

Figure 9.11 Depth predictions for pattern 5.

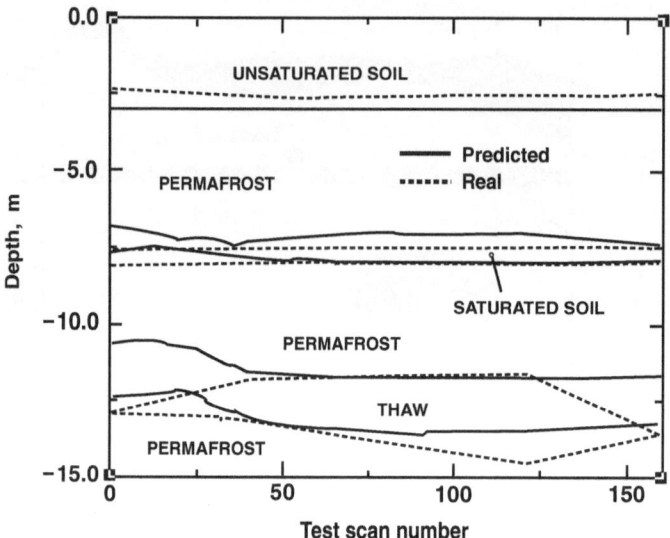

Figure 9.12 Depth predictions for pattern 6.

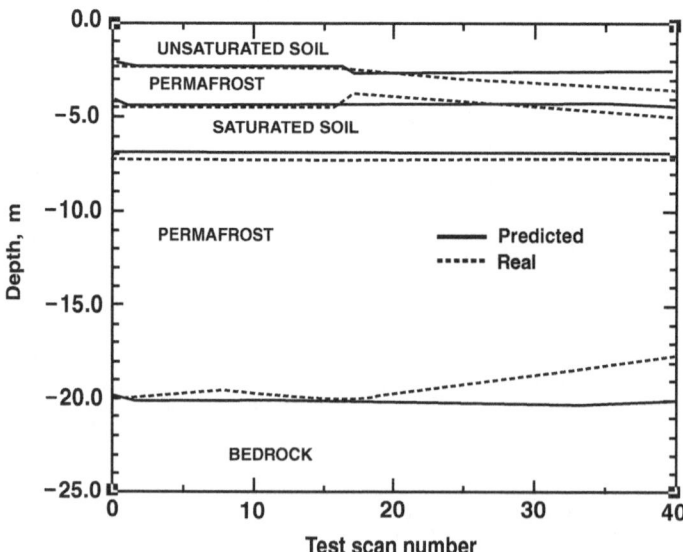

Figure 9.13 Depth predictions for pattern 7.

5, 6, and 7 (using corresponding NN2 modules) were predicted with good reliability as shown in Figures 9.11 to 9.13, respectively.

CONCLUSIONS

A novel approach to the decomposition of GPR data has been developed. It has a preprocessing section that performs an adaptive transform of the raw data such that only the salient features of the signal are retained. This reduction in input to the neural networks has reduced the training requirements of the system by two to three orders of magnitude. The neural networks are arranged in a hierarchical mode that reduces the uncertainty in each stage. The first neural network identifies successfully the subsurface pattern. The second neural network (one for each pattern) determines the various layer depths. Performance of each component was tested with synthetic as well as real GPR data and has demonstrated acceptable operation with respect to accuracy and error/noise tolerance.

ACKNOWLEDGMENTS

This work was funded in part by SERDP (Strategic Environmental Research and Development Program) Project 1049/6 cleanup thrust area and by CRREL (U.S. Army Cold Regions Research and Engineering Laboratory) SFRC Number DACA89-97-K0001.

REFERENCES

Arcone, S.E., D.E. Lawson, A.J. Delaney, J.C. Strasser, J.D. Strasser. Ground-penetrating radar reflection profiling of groundwater and bedrock in an area of discontinuous permafrost, *Geophysics*, 63(5), 1573–1584, 1998.

Haykin, S. *Neural Networks: A Comprehensive Foundation*, Macmillan College Publishing Company, New York, 1994.

Lawson, D.E., J.C. Strasser, J.D. Strasser, S.A. Arcone, A.J. Delaney, and C. Williams. Geological and geophysical investigations of the hydrogeology of Fort Wainwright, Alaska, Part I: Canol road area, CRREL Report 96–4, Cold Regions Research and Engineering Laboratory, Hanover, NH, 1996.

Parkhomenko, E.I. *Electrical Properties of Rocks*, Plenum Press, New York, 1967.

Plunkett, J. ECE Department, Worcester Polytechnic Institute, personal communications, 1997.

Repin, D.V. A Hierarchical Neural Network Based Data Processing System for Ground-Penetrating Radar, M.S. thesis, ECE Department, Worcester Polytechnic Institute, 1998.

Smithsonian Institution, Smithsonian Physical Tables, 9th ed., Washington, D.C., 1954.

Case Study Models for Cold Regions

CHAPTER 10

Permeability of Frozen Silt to Organic Contaminants

Kim Winnicky*

CONTENTS

INTRODUCTION

Permafrost is commonly considered to be impermeable to contaminants. In arctic regions, permafrost is therefore often relied on as a means of waste and spill containment (Kellems et al., 1991) and northern remediation programs are often limited to the active layer. Frozen ground, however, has a measurable hydraulic conductivity at temperatures several degrees below 0°C. At subzero temperatures, unfrozen water present in frozen soil migrates along thermal, osmotic, and pressure gradients (Burt and Williams, 1976; Horiguchi and Miller, 1983; Smith, 1985). Solutes present within soil water are subjected to similar forces and may be transported through and with the unfrozen water (Murrmann, 1973).

---

* This work was conducted in partial fulfillment of a master's degree at Carleton University. Research was conducted at the Geotechnical Science Laboratories, Ottawa, Ontario, Canada.

Although the coexistence of ice and liquid water in soils at temperatures below 0°C was described in the early 1950s, few studies have addressed the effects of unfrozen water in permafrost on contaminant containment (Mohammed et al., 1995). Several field studies have investigated the impacts of surficial petroleum spills on the biotic environment and on the flow of oil through the active layer; however, reports of the rate and extent of contaminant spreading are generally descriptive and limited to short-term migration through the active layer (Freedman and Hutchinson, 1975; Everett, 1978; Collins, 1983; Kershaw, 1990). Laboratory experiments addressing contaminant behavior in frozen soils are even more sparse, although two recent laboratory studies have investigated the behavior of contaminants in freezing soils (Tumeo and Davidson, 1993; Mohammed et al., 1995).

Permafrost is defined as ground (i.e., soil or rock) that remains at a temperature below 0°C for at least 2 years (Harris et al., 1988). According to this definition, permafrost is not a homogenous medium as it may be composed of soils of varying textures and moisture contents. Air-filled pores of nonsaturated frozen soil may permit the transport of both water-miscible and -immiscible liquids. Fissures and cracks present in permafrost may act as conduits to fluids, especially those with low freezing points and low viscosities at subzero temperatures.

In theory, capillary forces, diffusion, and advective flow may permit the transport of miscible and immiscible fluids through frozen ground. Research is therefore required to assess whether permafrost is, in fact, impermeable to contaminants. This chapter summarizes an initial investigation into the *permeability* of frozen soil to organic *contaminants*.

The objectives of this study are as follows:

1. To determine if frozen silt is permeable to organic contaminants
2. To address the influence of soil moisture, soil temperature, and contaminant miscibility on the permeability of frozen soil to contaminants

CASE STUDY

Macroscale studies reported herein were conducted at the Centre de Géomorphologie, Centre National de la Recherche Scientifique, Caen, France. A follow-up series of benchscale studies were conducted at the Geotechnical Science Laboratories at Carleton University, Ottawa, Canada (Winnicky, 1995). Macroscale studies investigated the fate of two petroleum hydrocarbons (arctic diesel fuel and a synthetic lubricating oil) injected into frozen ground.

The Station de Gel is a controlled environment facility located at the Centre National de la Recherche Scientifique. The facility is a refrigerated hall 18 m long × 8 m wide × 5 m high, with two adjacent rooms that accommodate instrumentation and mechanical equipment. The facility was filled with Caen silt to a depth of 240 cm and compacted to an average dry bulk density of 1.56 g cm^{-3}. A network of sensors distributed symmetrically through the facility monitored soil and air temperatures. Twenty-five YSI-44033 interchangeable thermistors and 160 copper-constantan thermocouples measured soil temperatures to depths of 185 cm at intervals ranging from 15 to 20 cm. Air temperatures were monitored by thermocouples 100 cm above the soil surface. Temperature data were collected every 12 hours over the course of the study.

Through the manipulation of air temperatures within the hall, two freeze–thaw cycles were induced over the 73-day study period, with an average active layer depth of 50 cm. A description of the layout and instrumentation is detailed in Geotechnical Science Laboratories (1993).

Caen silt is a homogenous soil composed of 3 to 10% sand, 78 to 85% silt, and 10 to 20% clay. It has a hydraulic conductivity ranging from 1×10^{-9} to 1.5×10^{-8} m s^{-1}. During the study, soil moisture contents ranged from 12% at the ground surface to 32% at 15 cm depth.

At the onset of the study, the ground was thoroughly frozen. A caterpillar-mounted "Sedi-Tech" geotechnical drill was used to bore two 5-cm-diameter holes into the frozen silt to 80 cm depth. Forty-eight hours following drilling, 300 mL of contaminant (arctic diesel fuel or lubricating oil) equilibrated to air temperature ($-8°C$) was poured into each of the holes.

Fluid levels within the boreholes (hereafter referred to as borehole levels) were monitored daily over the entire course of the experiment. No clearing was necessary in the diesel borehole, as all of the diesel had migrated from the borehole by day 25. To ease drilling, boreholes were then filled with saturated sand. The low temperature of the sand, which was saturated with a mixture of ice and water, minimized any potential melting.

The sand-filled boreholes were left to freeze for 1 week. On day 80, a single core was extracted from each of the boreholes. A 12-cm-diameter auger was used to extract a continuous core, which comprised both the frozen sand plug and the silts adjacent to the borehole. Each core was 90 cm in length, and soil was sampled to 10 cm below the bottom of the original hole. Eight samples were taken from each core and transported to Ottawa for further analysis. Gas chromatographic (GC) analysis was conducted at the Centre for Analytical and Environmental Chemistry at Carleton University.

RESULTS

Lubricating oil and arctic diesel fuel migrated laterally and vertically through the simulated permafrost. Little or no migration occurred until day 25 of the study (Figure 10.1). Infiltration of both contaminants began as soil temperatures warmed to $-0.1°C$ and continued even when soil temperature fell to $-0.4°C$.

In total, eight samples were extracted from the lubricating oil core. Figure 10.2 depicts soil concentrations in all samples extracted from the borehole. Maximum contamination (1.44×10^{-2} g g^{-1}) was detected 8 to 10 cm below the borehole, while minimum oil concentrations (1.74×10^{-3} g g^{-1}) were detected in the soil adjacent to the borehole 60 to 66 cm below the ground surface (Figure 10.2). Lateral migration had progressed a minimum of 8 cm over the 73-day study period. Lubricating oil penetrated a minimum of 14 cm into the silt underlying the borehole, with a sharp increase in oil concentration observed 8 cm below the borehole.

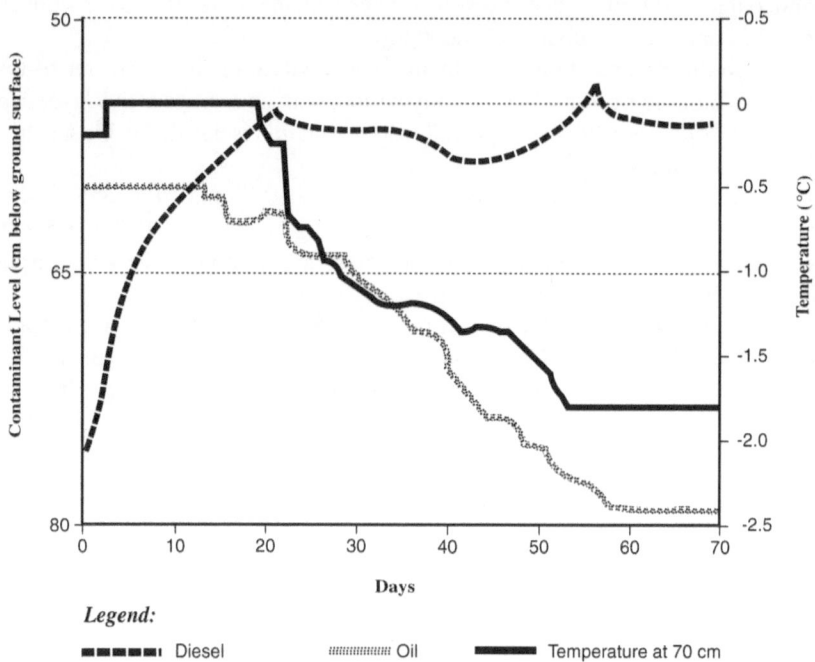

Figure 10.1 Contaminant levels and soil temperature over study period.

The behavior of diesel fuel was similar to that of lubricating oil. Diesel fuel was detected in all extracted samples, including one from the region above the original contaminant level (Figure 10.3). A concentration of 1.12×10^{-2} g g^{-1} was detected 10 to 20 cm above the contaminant source. Such contamination may have resulted from accidental splashing as the DF-A was poured into the borehole, though precautions were taken to prevent such inadvertent contamination. It is also possible that diesel fuel migrated toward the surface through nonsaturated pores in response to capillary and osmotic forces as either a liquid or vapor.

Contamination levels above the source were an order of magnitude greater than those detected in soil immediately adjacent to or below the boreholes. Diesel fuel penetrated a significant distance into the soil adjacent to and underlying the borehole. Diesel migrated a minimum of 8 cm laterally, with concentrations ranging from 8.7×10^{-3} g g^{-1} immediately adjacent to the borehole to 2.27×10^{-4} g g^{-1} at the most distal sample. Diesel penetrated a minimum of 5 cm into the silt beneath the borehole.

Frozen silt initially contained the diesel fuel. As soil temperatures approached $-0.1°C$ (day 25), diesel levels within the borehole declined and continued to fall until day 56 when the borehole was completely evacuated (Figure 10.1). At the conclusion of the study, borehole depth was reduced to approximately 75 cm. This reduction in depth was due to minimal sloughing from the borehole sides, which occurred during the initial days of the study.

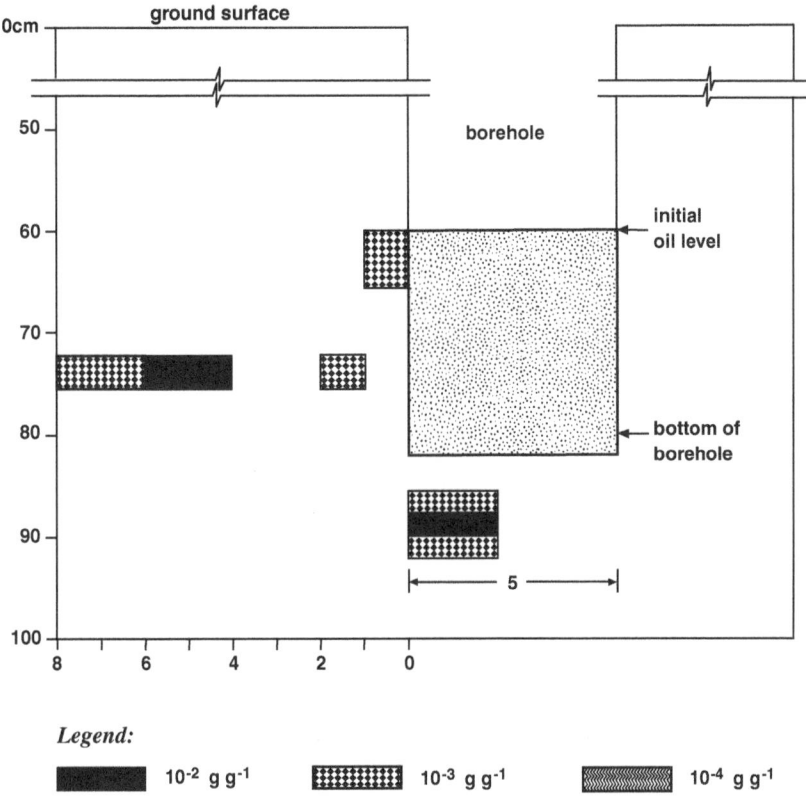

Figure 10.2 Oil concentrations (gg⁻¹) in soil adjacent to borehole.

METHODS OF TRANSPORT

In both cores, the extent and magnitude of contaminant migration were higher than anticipated given the immiscible nature of arctic diesel fuel and lubricating oil. The two most probable mechanisms of transport were the melting of ice by water-soluble fractions of the contaminants and subsequent aqueous transport, or movement through small fissures and along pressure gradients induced by soil warming.

The abrupt drop in borehole levels observed as soil temperatures warmed to $-0.1°C$ suggests that thermally induced soil structural changes and not the melting of pore ice were responsible for the increased soil permeability. Soil structure is altered in response to mineral and pore volume changes generated by soil warming and cooling, and by the migration of unfrozen moisture along pressure, osmotic, and thermal gradients. During freezing, the conversion of water to ice results in a 9% volumetric increase. As soil cools, its pore volume increases by an amount depending on the volume of water converted to ice, which in turn is dependent

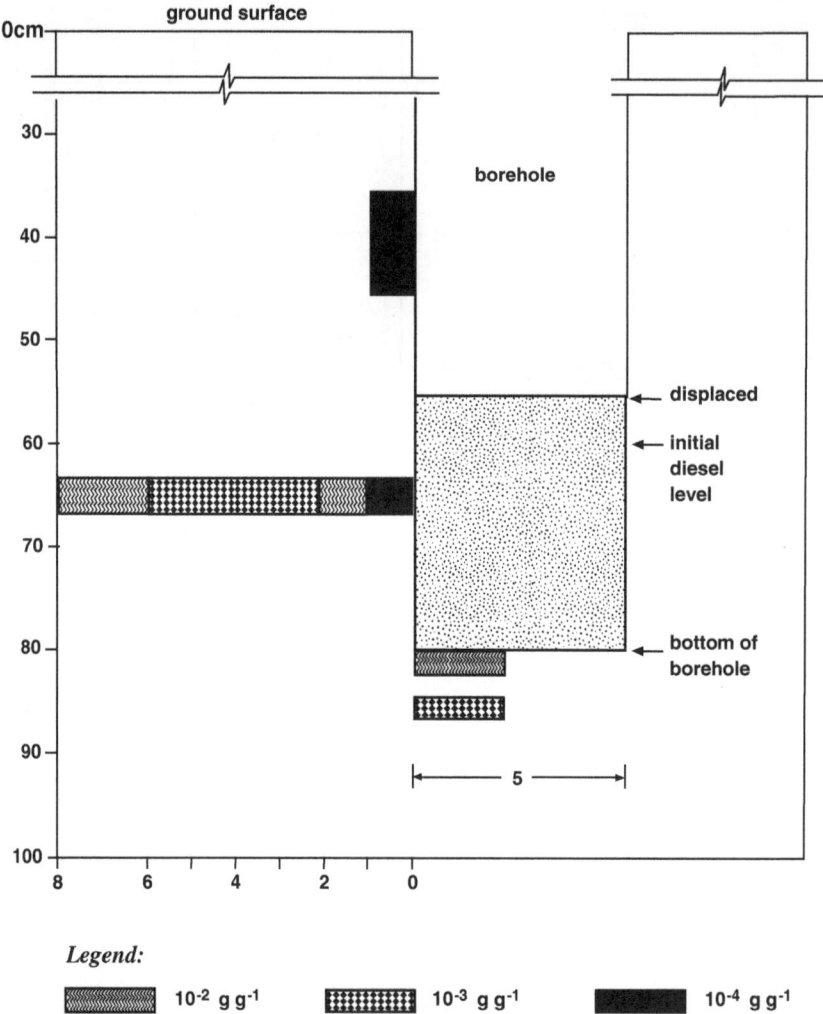

Figure 10.3 Diesel concentrations (gg⁻¹) in soil adjacent to borehole.

upon the soil-freezing characteristic curve. Soil cooling also leads to a contraction of the mineral particles. The magnitude of mineral volumetric decrease is dependent upon the coefficient of thermal expansion, which is approximately $10^{-5}°C^{-1}$ for frozen soil.

As soil temperatures increase, the above processes are reversed. The mineral particles expand as dictated by the coefficient of thermal expansion, while pore phase volume decreases as a fraction of the soil ice is converted to liquid water. While there is an overall decrease in pore volume, soil particles remain in place as they are supported and separated by the ice matrix. The decrease in pore phase volume leads to a fall in pore pressure, which induces fluid infiltration and transport through the soil along these gradients. In the boreholes, such pressure gradients likely induce the infiltration of hydrocarbons into the proximate soil.

Following contamination, the frozen silt remained permeable to both contaminants even as soil temperatures fell to –0.4°C. As a fluid infiltrates a porous matrix, it establishes continuous flow paths. The organic liquid flow paths established near –0.1°C would remain liquid and mobile at lower temperatures due to the low pour points of the fluids. The presence of such conduits would account for the continued flow in contaminants in the borehole even as soil temperatures decreased.

SUMMARY

At temperatures near 0°C, frozen silt was permeable to lubricating oil and diesel fuel. While frozen soil initially contained the petroleum mixtures, contaminant migration was initiated as soil temperatures at depth warmed to –0.1°C and continued even as temperatures approached –0.4°C. Given the immiscible nature of the fluids employed, the main transport mechanism is considered to be transport under thermally induced negative pressures. Once established, continuous contaminant flow paths appear to have remained liquid and mobile even as the soil cooled.

The results presented indicate that permafrost cannot be considered impermeable, as contaminants penetrate frozen soil under certain conditions. It must be emphasized that this study was limited to soil temperatures near 0°C. Bench-scale studies reported in Winnicky (1995) indicate that frozen silt is permeable to water-soluble substances at temperatures as low as –10°C, and that nonsaturated frozen soil is permeable to both water-miscible and -immiscible liquids. Further studies are required to define the permeabilities of soil at a variety of saturations, textures, and temperatures to a number of contaminants. Such experimental data could assist in increasing the predictive capacity of models addressing the fate and contaminants in cold regions.

ACKNOWLEDGMENTS

Financial support for this study was generously provided by the Canadian Department of National Defence, the Northern Science Training Program, and the National Energy Board. I gratefully acknowledge the support and guidance of Dr. Peter Williams, Dr. Chris Burn, and Dan Riseborough, all of Carleton University.

REFERENCES

Burt, T.P. and P.J. Williams. Hydraulic Conductivity in Frozen Soils. *Earth Surface Processes.* 1(3), 349–360, 1976.

Collins, C.M. Long-Term Active Layer Effects of Crude Oil Spills in Interior Alaska. In *Proceedings of the Fourth International Conference on Permafrost, Fairbanks, Alaska. Vol. 1.* National Academy of Sciences, Washington, D.C., 175–179, 1983.

Everett, K.R. Some Effects of Oil on the Physical and Chemical Characteristics of Wet Tundra Soils. *Arctic*. 31, 260–276, 1978.

Freedman, W. and T.C. Hutchinson. Physical and Biological Effects of Experimental Crude Oil Spills on Low Arctic Tundra in the Vicinity of Tuktoyaktuk, N.W.T., Canada. *Canadian Journal of Botany*. 54, 2219–2230, 1975.

Geotechnical Science Laboratories. *Gas Pipelines, Oil Pipelines and Civil Engineering in Arctic Climates*. Proceedings of a Seminar in Caen, France. Geotechnical Science Laboratories, Carleton University, Ottawa, 1993, 173 pp.

Harris, S.A., H.M. French, J.A. Heginbottom, G.H. Johnston, D.C. Ladanyi, and R.O. van Everdingen. *Glossary of Permafrost and Related Ground-Ice Terms*. Technical Memorandum No. 1422. Permafrost Subcommittee, Associate Committee on Geotechnical Research, National Research Council of Canada, Canada, 1988.

Horiguchi, K. and R.D. Miller. Hydraulic Conductivity Functions of Frozen Materials. In *Proceedings of the Fourth International Conference on Permafrost, Fairbanks, Alaska*. National Academy Press, Washington, D.C., 504–508, 1983.

Kellems, B.L., R.W. Slocum, and M.C. Kavanaugh. Alternatives for Closure of Solid Oily Waste Sites on the North Slope of Alaska. In *International Arctic Technology Conference, Fairbanks, Alaska*. Society of Petroleum Engineers, 155–169, 1991.

Kershaw, G.P. Movement of Crude Oil in an Experimental Oil Spill on the SEEDS Simulated Pipeline Right-of-Way, Fort Norman, NT, *Arctic*. 43(2), 176–183, 1990.

Mohammed, A.M., R.N. Yong, and M.T. Mazus. Contaminant Migration in Engineered Clay Barriers Due to Heat and Moisture Redistribution Under Freezing Conditions. *Canadian Geotechnical Journal*. 32, 40–59, 1995.

Murrmann, P.P. Ionic Mobility in Permafrost. In *Proceedings of the Second International Conference on Permafrost, Yakutsk, USSR*. North American Contribution. National Academy of Sciences, Washington, D.C., 352–358, 1973.

Smith, M.W. Observations of Soil Heaving and Frost Heaving at Inuvik, Northwest Territories, Canada. *Canadian Journal of Earth Sciences*. 22(3), 283–290, 1985.

Tumeo, M.A. and B. Davidson. Hydrocarbon Exclusion from Groundwater During Freezing. *Journal of Environmental Engineering*. 119(4), 715–724, 1993.

Winnicky, K.L. On the Permeability of Frozen Silt to Organic Contaminants. Master's thesis, Carleton University, Ottawa, Ontario, 1995.

CHAPTER 11

A Pore-Scale Model for Soil Freezing

Lin A. Ferrand

CONTENTS

INTRODUCTION

Predictions of moisture flux and associated contaminant transport in soils that undergo seasonal freeze–thaw cycles require that a series of complex, material-dependent relationships be determined. These include relationships between (1) fluid pressures and the amount of each fluid present; (2) the amount of each fluid present and the permeability of the soil to each; and (3) temperature and the amount of liquid water present. The first is generally expressed as a relationship between capillary pressure (P_c) and saturation or moisture content (S_w or θ_w) and is sometimes referred to as the soil water characteristic (SWC). (P_c is defined as the pressure in the fluid that preferentially wets the soil grains subtracted from the pressure in the other fluid, S_w is the volume of water per volume of voids, and θ_w is the product of S_w and the soil porosity.) The third is expressed as a relationship between temperature below 0°C or freezing temperature depression (ΔT) and liquid water saturation or liquid moisture content (S_L or θ_L), where S_L is the volume of liquid water per volume of voids and θ_L is the product of S_L and porosity. This is referred to as the soil freezing characteristic (SFC). These relationships all depend upon the geometry and topology of the soil pore space and are interdependent. In the work reported here, we apply a pore-scale computational model that simulates the SWC and the SFC, with emphasis on the soil freezing results. Results of SWC simulations can be found in Ferrand and Celia (1992), Ferrand et al. (1994), and Ferrand and Sulayman (1996).

In a typical laboratory SFC experiment, a water-saturated soil sample (Figure 11.1A) is subjected to a series of incrementally decreasing temperatures. After equilibrating at a given temperature, the fraction of water that remains in the liquid state is determined. Methods that have been used to make this determination include calorimetry (Williams, 1964), pulsed nuclear magnetic resonance (Black and Tice, 1989), and electrical conductance measurement (Gunnink and El-Jayyousi, 1993). The computational freezing model mimics this experiment.

THE COMPUTATIONAL MODEL

The percolation-based model reproduces continuum-scale phenomena (such as the SWC and SFC) by simulating pore-scale events (displacement of water by air or liquid water by ice) in an interconnected lattice of many pores. In the computational model, soil pore space is idealized as a regular cubic lattice of spherical pore bodies connnected by cylindrical pore throats (Figure 11.1B). Pore radii are assigned from specified probability distributions. Figure 11.2 shows segments of cross sections through cubic lattices that illustrate two versions of the internal structure of pore space. In the first (type A), pore throat and pore body radii are assigned from separate pore size distributions. Pore body radii that are (1) larger than one half the distance between pore body centers or (2) smaller than any of the connecting pore throat radii are refused by the lattice generating algorithms. In the second (type B), pore bodies are given the radius of the largest connecting pore throat.

In a typical (saturated) SFC simulation, the lattice is initially saturated with a single fluid. Like the laboratory sample, the lattice is in contact with external reservoirs of the initially resident phase (liquid water in the simulations reported here) and the invading phase (air or ice). It is subjected to a series of incremental pressure or temperature changes and allowed to equilibrate at each. That is, phase interfaces are allowed to move through connected pores until they reach stable configurations. When this is completed, the lattice is scanned to determine what fraction of pore space remains water filled.

The displacement algorithm may require that wetting fluid be hydraulically connected (through a continuous pathway of wetting fluid-filled pores) to its external reservoir in order to drain. If this capillary trapping rule is applied, then the final state of a drained lattice is a nonzero wetting fluid saturation. If not, the lattice drains completely. For the air–water case, Dullien et al. (1989) found that true capillary trapping during drainage only occurs in porous materials with very smooth surfaces.

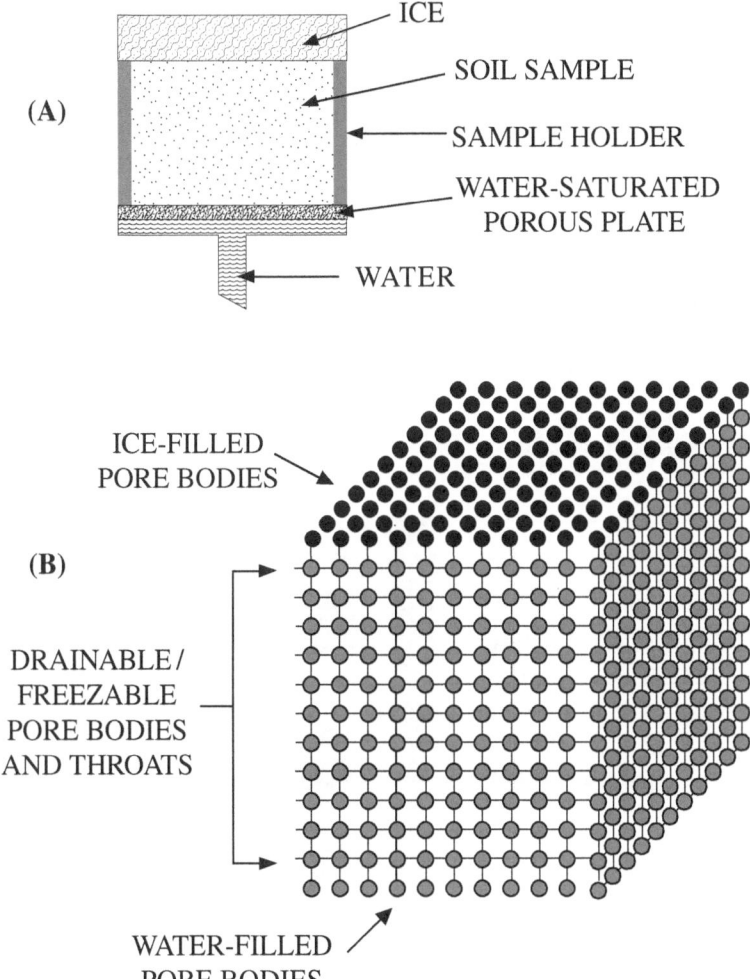

Figure 11.1 Schematics of a freezing cell and a computational lattice. (A) Schematic of a freezing cell for laboratory determination of the soil freezing characteristic (SFC); (B) schematic of a computational lattice used to simulate the laboratory SFC experiment.

For rough grains they observed that wetting fluid maintains hydraulic connection through wetting fluid-filled microchannels on solid surfaces. While flow through these channels may be extremely slow, nearly all wetting fluid was found to be extractable given sufficient equilibrium time. There appears to be less ambiguity in the ice–water case since freezing does not seem to require hydraulic connection of the remaining fluid.

Implementation of a lattice displacement algorithm requires specification of a criterion for interface stability. That is, a relationship between the radius of a pore that can support an interface and the independent variable of the simulation is needed.

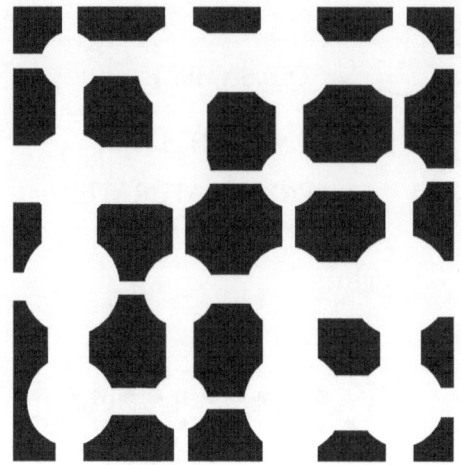

**TYPE A: PORE BODY
RADII GENERATED
FROM INDEPENDENT
PORE SIZE
DISTRIBUTION
FUNCTION**

**TYPE B: PORE BODY
RADIUS = RADIUS
OF LARGEST
CONNECTING PORE
THROAT**

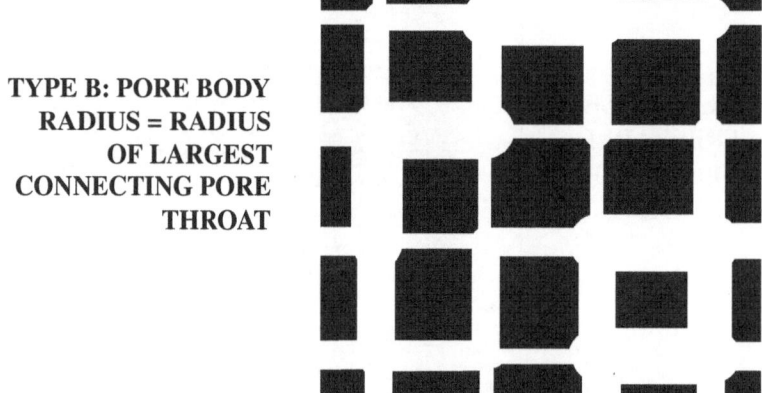

Figure 11.2 Magnified views of sections through three-dimensional pore lattices.

The model was originally developed for soil drainage and imbibition simulations. In this case, the Young–Laplace equation serves as the stability criterion:

$$P_c = 2\sigma_{\alpha\beta}\cos\theta_{\alpha\beta}\left(\frac{1}{R}\right) \tag{11.1}$$

where P_c is capillary pressure (the independent variable), $\sigma_{\alpha\beta}$ is fluid/fluid interfacial tension, $\theta_{\alpha\beta}$ is fluid/fluid/solid contact angle, and R is (circular) pore radius. (Computational models of this type have been used by many researchers to investigate phenomena associated with fluid displacement and flow in porous media. Celia et al. [1995] provide a comprehensive review of recent applications of pore-scale models to hydrology-related problems.) For example, during standard drainage of water from a hydrophilic soil, water/air interfaces are displaced through connecting pores until they reach a pore with a radius smaller than $2\sigma_{aw}\cos\theta_{aw}$ divided by the currently imposed capillary pressure.

Freezing Displacement Rule

Adaptation of the lattice model for soil freezing simulations requires specification of a criterion analogous to Equation 11.1 for ice/water interfaces. A number of alternative criteria, based on work reported in the soils literature, are described below.

It may be reasonable to assume, for the case of a wetting phase being displaced by a nonwetting phase in a given porous medium, that the spatial distribution of the wetting phase (at a given value of wetting fluid saturation) is independent of the properties of the nonwetting fluid. For example, given a water-wet soil, the distribution of liquid water in the pore space at a given value of liquid water content will be the same whether the soil was undergoing drainage (displacement of liquid water by air) or freezing (displacement of liquid water by ice). Thus, one approach to defining an ice/water displacement rule is to relate capillary pressure values during drainage to temperature values during freezing to a series of water contents ($\theta^* = \theta_w = \theta_L$) for the same soil.

An early example of this comparison, made in an experimental context, can be found in Williams (1964), who gave this relation as

$$P_c(\text{cm H}_2\text{O}) = \frac{h}{(T_o + 273.16)g}\Delta T(^\circ\text{C}) \qquad (11.2)$$

where h is the specific heat of fusion for the aqueous phase, T_o is the bulk freezing temperature of the aqueous phase, g is the acceleration of gravity, and ΔT is the freezing point depression. This expression was given without derivation and was referred to Schofield (1935) and the earlier work of Poynting (1881) and Hudson (1906). Williams (1964) performed drainage and freezing experiments for several clay soils, a silt, and an iron ore and found good agreement between his experimental data and Equation 11.2. Substituting $h = 3.333 \times 10^5$ J/kg, $g = 9.81$ m/s^2, and $T_o = 0^\circ$C in Equation 11.2 and making the necessary unit conversions, we find P_c (cm H$_2$O) = 12,499 ΔT ($^\circ$C).

Koopmans and Miller (1966) noted that quantitative comparisons between soil freezing and soil moisture curves are possible only for soils that belong to one of two extreme types: granular soils in which solid grains are always in contact with each other and colloidal soils in which solids are always separated by liquid films. They found that scaled soil freezing curves for granular soils closely matched soil moisture curves at relatively high water/liquid water saturations. At lower saturations liquid water contents during freezing were consistently lower than water contents during drainage. They were able to eliminate a great deal of this discrepancy by repeating drainage experiments with significantly longer equilibration times. This observation and the results of Dullien et al. (1989) fuel the continuing debate on the use of capillary trapping in lattice simulations of wetting fluid displacement.

Koopmans and Miller (1966) give two different expressions relating P_c and ΔT at the same liquid water content. For colloidal soils

$$P_c(\text{cm H}_2\text{O}) = \frac{h}{V_i(T_o + 273.16)}\Delta T(^\circ\text{C})\tag{11.3}$$

where V_i is the specific volume of ice, while for granular soils

$$P_c(\text{cm H}_2\text{O}) = \frac{\sigma_{aw}}{\sigma_{iw}}\frac{h}{V_i(T_o + 273.16)}\Delta T(^\circ\text{C})\tag{11,4}$$

where σ_{iw} is the ice/water interfacial tension. Note that the expression for granular soils in Equation 11.4 depends upon fluid interfacial properties, while the expression for colloidal soils in Equation 11.3 does not. Substituting V_i in Equation 11.3 gives P_c (cm H$_2$O) = 11,315 ΔT (°C) for colloidal soils. Koopmans and Miller (1966) performed drainage and freezing experiments on a montmorillonite clay soil and found good agreement between their experimental results and Equation 11.3. Note that this proportionality constant is close to the result of Williams (1964) whose experimental soils were mostly clays.

Lacking an independent measurement for σ_{iw}, Koopmans and Miller (1966) fit in Equation 11.4 to experimental drainage and freezing data for 2 to 4 and 4 to 8 μm silt fractions to find $\sigma_{aw}/\sigma_{iw} = 2.2$. Substituting this value back in Equation 11.4 gives P_c (cm H$_2$O) = 24,873 ΔT (°C) for granular soils.

Brun et al. (1977) give an expression for freezing point depression in individual pores as

$$R = \frac{65.67}{\Delta T} + 0.57\tag{11.5}$$

where R is pore radius in nanometers. For the sake of comparison, we combine this expression with the Young–Laplace Equation 11.1 which gives

$$P_c(\text{cm H}_2\text{O}) = \frac{2 \times 10^6 \sigma_{aw}\cos\theta}{\rho g(-64.67 + 0.57\Delta T)}\Delta T(^\circ\text{C})\tag{11.6}$$

where ρ is the density of water. While this expression is slightly nonlinear, a good approximation for a perfectly water–wet soil ($\theta = 0$) is given by P_c (cm H$_2$O) = 22,935 ΔT (°C).

Gunnink and El-Jayyousi (1993) give another expression for freezing point depression in pores as

$$\Delta T = \frac{2\sigma_{iw}(T_o + 273.16)}{\rho h R}\tag{11.7}$$

They estimate ice/water interfacial tension as

$$\sigma_{iw} = \sigma_{iw(o)} + k\Delta T \qquad (11.8)$$

where $\sigma_{iw(o)}$ is 2.9×10^{-2} J/m^2 and k is 2.5×10^{-4} J/m^2 °C (referred to Hesstvedt, 1964). Combining Equation 11.6 with 11.1 gives

$$P_c(\text{cm H}_2\text{O}) = \frac{\sigma_{aw}\cos\theta\rho h}{\sigma_{iw}(T_o + 273.16)} \qquad (11.9)$$

Like Equation 11.5, this expression is very slightly nonlinear due to the dependence of σ_{iw} on ΔT. However, a good approximation for a perfectly water–wet soil is given by P_c (cm H$_2$O) = 31,202 ΔT (°C).

These results are summarized in Figure 11.3, which is a plot of freezing point depression vs. capillary pressure for identical values of θ^*. The upper curves (open symbols) represent relationships between these variables for colloidal soils, while

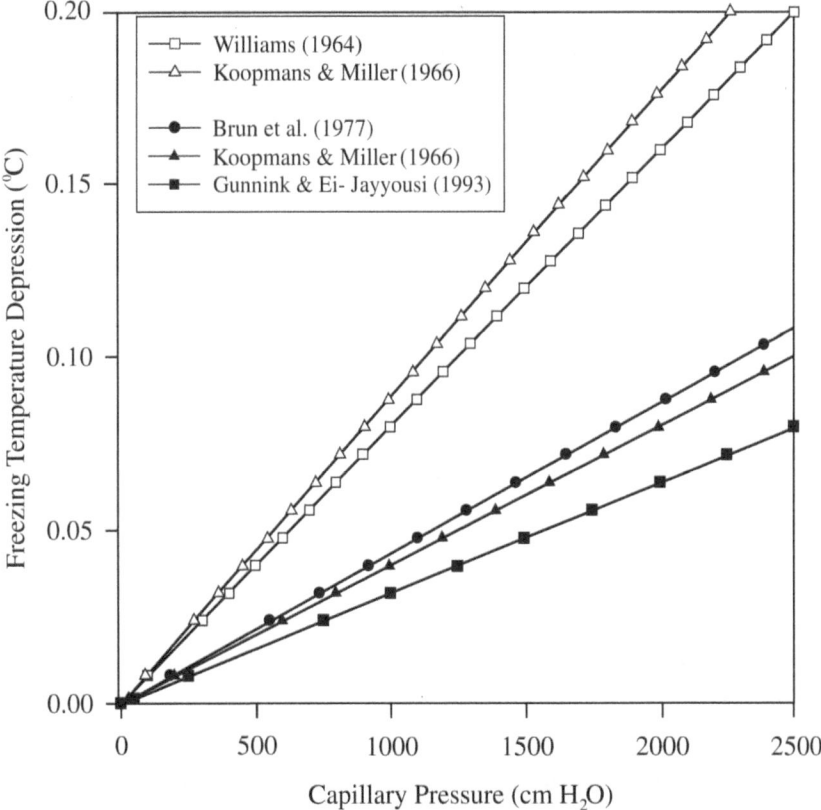

Figure 11.3 Published formulations for the relationship between ΔT and P_c at the same value of liquid moisture content.

the lower curves (closed symbols) represent relationships for granular soils. Black and Tice (1989) note that in colloidal soils, the distribution of the aqueous phase within void space is dominated by adsorption to solid surfaces, while in granular soils this spatial distribution is controlled by capillary forces. The pore-scale lattice model described herein simulates displacement of one fluid by another as a capillary phenomenon. It is, therefore, unable to reproduce the behavior of colloidal soils in its present form. Further discussion will be limited to granular soil behavior. Since it allows maximum flexibility in defining fluid parameters, we adopt Equation 11.6 as our water/ice displacement rule for the purposes of the current model.

COMPUTATIONAL RESULTS

Freezing of Saturated Soils

Drainage (without capillary trapping) and freezing experiments were performed on three computational soils. In every case, the lattice was completely filled with liquid water to initiate the simulation. Type A lattices (see Figure 11.2) were used. Soil types include homogeneous (open circles), fine soil with coarse lenses (open squares), and coarse soil with fine lenses (open triangles) (Ferrand and Celia, 1992). Earlier work by the author and others (Rajaram et al., 1997) suggests that, for nonhomogeneous cases, lattices at least 12 correlation scales long in each direction are required. Since the lens size in these simulations was $3 \times 3 \times 3$ bond lengths, $36 \times 36 \times 36$ lattices were used.

Results are shown in Figure 11.4, which is a plot of freezing temperature depression vs. capillary pressure of approximately the same values of liquid water content ($\theta_w = \theta_L$). The experimental results of Koopmans and Miller (1966) for granular soils (solid diamonds) and the curves given by Koopmans and Miller (1966) and Gunnink and El-Jayyousi (1993) formulations are included for comparison. (Recall that the slope of the Koopmans and Miller formulation was found by calibration to the data shown and that the computational model uses the Gunnink and El-Jayyousi formulation.)

Freezing of Unsaturated Soils

Given appropriate displacement rules, drainage and freezing algorithms can be combined in a single computational model to simulate the freezing of partially water-saturated soils. For the simulations reported here, it was assumed that the presence of air/water interfaces in the system does not affect the displacement of ice/water interfaces. Additional analysis will be required to establish the magnitude of the error resulting from this assumption.

The combined model can be used to generate a family of freezing curves, each with a different initial liquid moisture content, for a given soil. An example is shown in Figure 11.5, which is a plot of ΔT vs. θ_L for varying initial moisture content. The curves shown were generated using a homogeneous uncorrelated

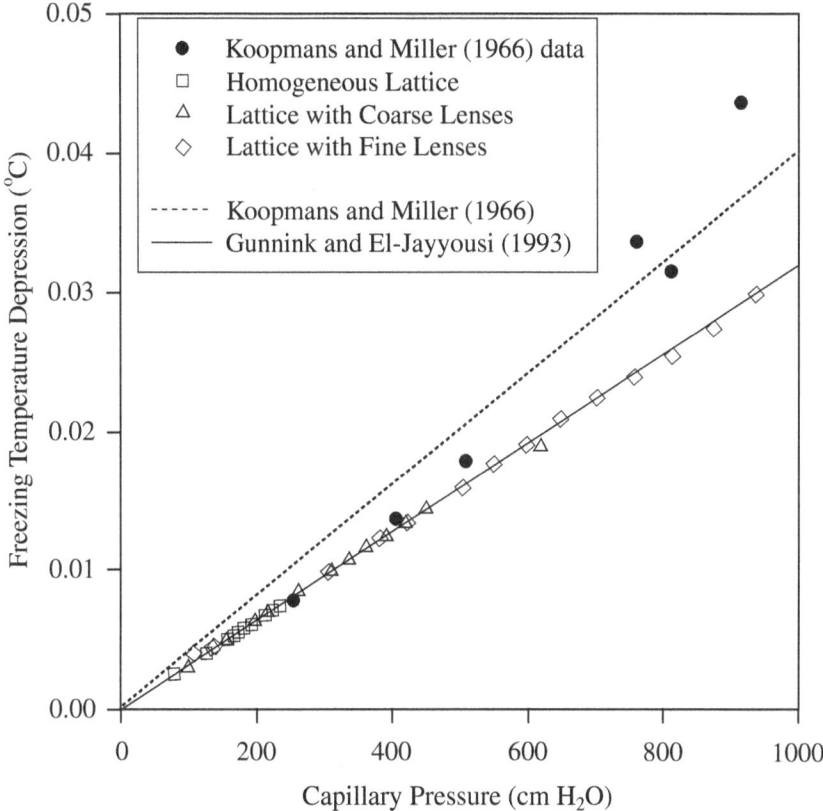

Figure 11.4 Correlation between ΔT and P_c at the same value of liquid moisture content: homogeneous uncorrelated (simulated) soil (open circles), fine (simulated) soil with coarse lenses (open squares), coarse (simulated) soil with fine lenses (open triangles), Koopmans and Miller (1966) granular soil data (diamonds), Equation 11.9 (solid line), Equation 11.4 (dotted line).

(type B) lattice. Each simulation was started with the lattice fully water-saturated ($P_c = 0.0$ cm H_2O, $\Delta T = 0.0°C$, $\theta_{sat} = 0.232$). The saturated freezing curve (circles) was generated by imposing incrementally decreasing temperature. All other simulations started with incremental increases in capillary pressure (until the desired value of initial water content for freezing was reached) followed by incremental temperature decreases.

While we believe that this simulation result qualitatively represents the behavior of granular soils, we have been unable to find laboratory data for comparison. The single data set currently available to the author (unpublished data from the U.S. Army Cold Regions Research and Engineering Laboratory for Hecla [North Dakota] soil) does not seem to contradict this result. However, there is insufficient resolution in the data between initial and final values of liquid moisture content to provide confirmation. In addition, the experimental data suggest that an irreducible liquid moisture content may be reached during the freezing process and that this may be independent of the initial moisture content. The current SFC model does not include

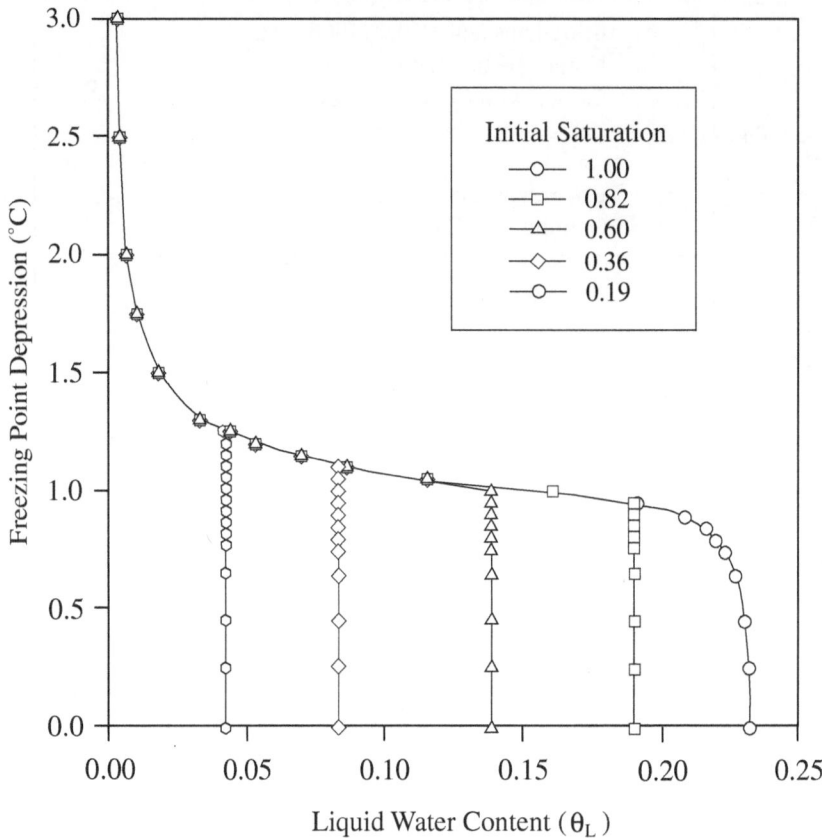

Figure 11.5 Lattice model-generated relationship between ΔT and θ_L for varying initial moisture content.

a mechanism, perhaps analogous to the capillary trapping mechanism in the SWC model, that might identify some fraction of water as essentially unfreezable.

DISCUSSION

The work reported above demonstrates the ability of the pore-scale model to reproduce multiphase displacement phenomena such as soil drainage and freezing. As many researchers have demonstrated, the primary drainage and water-saturated freezing curves can sometimes be related by a straightforward scaling based on interfacial properties. However, meaningful predictions of moisture flux and contaminant transport under field conditions in cold regions require a more complex set of relationships. The freezing behavior of partially saturated soils is a fairly simple example. The long-term goal of the research described herein is to develop an alternative to exhaustive laboratory characterization of transport-related properties of soils undergoing freeze/thaw cycles. Demonstration that the model can reproduce simple and relatively easily measurable relationships is a necessary step in that development.

ACKNOWLEDGMENTS

This work was sponsored in part by the U.S. Department of the Army, Environmental Quality Technology Installation, Restoration Project BT25-EC-B08 entitled "Pore-Scale Modeling of Multiphase Contaminant Transport in Freezing and Thawing Soils." The author acknowledges Steve Grant and John M. Cullinane. The views and conclusions contained herein are those of the author and should not be interpreted as necessarily representing the official policies or endorsements, either expressed or implied, of the Department of the Army or the U.S. Government. The author wishes to thank the anonymous technical reviewer for his/her valuable suggestions.

REFERENCES

Black, P.B. and A.R. Tice, Comparison of Soil Freezing Curves and Soil Water Curve Data for Windsor Sandy Loam, *Water Resour. Res.*, 25(10), 2205–2210, 1989.

Brun, M., A. Lallemand, J.-F. Quinson, and C. Eyraud, A New Method for the Simultaneous Determination of the Size and Shape of Pores, the Thermoporometry, *Thermochim. Acta*, 21, 59–88, 1977.

Celia, M.A., P.C. Reeves, and L.A. Ferrand, Recent Advances in Pore-Scale Models for Multiphase Flow in Porous Media, *Rev. Geophys.*, Supplement, 1049–1057, 1995.

Dullien, F.A.L., C. Zarcone, I.F. MacDonald, A. Collins, and R.D.E. Bochard, The Effects of Surface Roughness on the Capillary Pressure Curves and the Heights of Capillary Rise in Glass Bead Packs, *J. Colloid Interface Sci.*, 127(2), 362–372, 1989.

Enustun, B.V., H.S. Senturk, and O. Yurdakul, Capillary Freezing and Melting, *J. Colloid Interface Sci.*, 65(3), 509–516, 1978.

Ferrand, L.A. and M.A. Celia, The Effect of Heterogeneity on the Capillary Pressure–Saturation Relation, *Water Resour. Res.*, 28(3), 859–870, 1992.

Ferrand, L.A., J.A. Sulayman, H. Rajaram, and P.C. Reeves, Calibration of a Pore-Scale Network Model for Unsaturated Soils, in *Proc. 14th Annual Hydrology Days*, Hydrology Days Publishing, Atherton, CA, 1994.

Ferrand, L.A. and J.A. Sulayman, A Computational Investigation of Some Effects of Temperature on Soil Moisture, *Water Resour. Res.*, 32(12), 3429–3436, 1996.

Gunnink, B.W. and J. El-Jayyousi, Soil Fabric Measurement Using Phase Transitition Porosimetry, *J. Geotech. Eng.*, 119(6), 1019–1036, 1993.

Hesstvedt, E., The Interfacial Energy Ice/Water, *Publ. No. 56, Norwegian Geotechnical Inst.*, Oslo, Norway, 7–10, 1964.

Hudson, C.S., The Freezing of Pure Liquids and Solutions Under Various Kinds of Positive and Negative Pressures and the Similarity Between Osmotic and Negative Pressure, *Phys. Rev.*, 22, 257–264, 1906.

Koopmans, R.W.R. and R.D. Miller, Soil Freezing and Soil Water Characteristic Curves, *Soil Sci. Soc. Am. Proc.*, 30, 680–685, 1966.

Poynting, J.H., Change of State: Solid-Liquid, *Philos. Mag.*, 5(12), 32–48, 1881.

Rajaram, H., L.A. Ferrand, and M.A. Celia, Prediction of Relative Permeabilities for Unconsolidated Soils Using Pore-Scale Network Models, *Water Resour. Res.*, 33(1), 43–52, 1997.

Schofield, R.K., The pF of Water in Soil, *3rd Int. Congr. Soil Sci.*, 2, 37–48; 3, 182–186, 1935.

Williams, P.J., Unfrozen Water Content of Frozen Soils and Soil Moisture Suction, *Geotechnique*, 14(3), 231–246, 1964.

CHAPTER 12

Sensible Heat Exchange during Snowmelt

John M. Baker, Gerald N. Flerchinger, and Egbert J.A. Spaans

CONTENTS

INTRODUCTION

In many parts of the world snow is a regular feature of the annual weather cycle, and its ultimate disposition through evaporation and melt can have profound impacts both locally and regionally. Surface and subsurface flow of meltwater can cause disastrous flooding, particularly when supplemented with spring rains. Snowmelt can be vital in replenishing soil moisture depleted by crops during the previous growing season, and deeper percolation is a contributor to aquifer recharge. Snow and snowmelt also exert significant effect on the soil thermal regime, with potentially important consequences for early season plant growth and development. Environmental impacts associated with snowmelt include sediment and chemical loading of surface waters due to runoff. Finally, regional weather patterns are influenced by surface/atmosphere energy exchange processes with snowpacks.

For all of these reasons, it is important to develop a predictive capability based on a physical understanding of snowmelt processes. Toward that end, we have developed and operated instrumentation systems for measuring energy, heat, and mass exchange in a farm field during winter and spring. The resulting data have been used to test an existing model of heat and mass transfer through a soil/snow system — the SHAW model, developed by Flerchinger and Saxton (1989). In this chapter we examine the ability of the SHAW model to estimate sensible heat exchange between the surface and the atmosphere during and following snowmelt.

FIELD MEASUREMENTS

The field research was conducted during the winter of 1993–1994 at the University of Minnesota Agricultural Experiment Station in Rosemount. It is located 20 km south of St. Paul (44° 45' N, 93° 05' W) on a relatively level glacial outwash plain. The soil at the site is a Waukegan silt loam (fine, mixed, mesic typic hapludoll), in which the surface horizons have a relatively high organic carbon content and a combined thickness of approximately 1 m. The underlying material is outwash sand and gravel, and the water table is approximately 20 m below the surface. Mean annual temperature at this site is 6.96°C, and mean annual precipitation is 823 mm, 17% of which falls during the winter months of December to March in a typical year. The mean number of days on which snow cover exceeds 2.5 cm is 87, with the average first and last dates of such cover being November 28 and March 27 (Kuehnast et al., 1982).

The field used for the experiment is 17 ha, with a semipermanent mast in the center that enjoys fetch in excess of 180 m in all directions. The field was plowed and disked on November 12, 1993, following harvest of maize. Data necessary for the model were measured at 15-s intervals and averaged every 30 min. These included incoming solar radiation, wind speed, air temperature, and relative humidity, all measured at 2 m above the soil surface, and precipitation measured with a tipping bucket and totaled every 30 min. Additional measurements included wind-speed at three other heights, air (and/or snow) temperatures at 12 heights above the surface from 0 to 50 cm, incoming and outgoing longwave radiation, incoming and reflected solar radiation, snow depth, and surface temperature by infrared thermometry. Soil temperatures and soil liquid water contents were measured hourly using thermistors and time domain reflectometry. Periodic manual measurements included total water content at 20-cm-depth increments by neutron moderation and snow water equivalent by gravimetric sampling.

Sensible heat flux was measured by eddy covariance, using a one-dimensional sonic anemometer and a fine-wire thermocouple (Campbell Scientific), mounted 2 m above the surface and sampled at 10 Hz. The mean turbulent flux of sensible heat is determined by measuring and computing the covariance of the vertical wind velocity and the "heat content" of the air:

$$H = C_p \overline{\rho_a} \overline{w'T'} \tag{12.1}$$

The flux, H, is given in W m^{-2}, C_p is the heat capacity of air in J kg^{-1}, ρ_a is the mean density of dry air in kg m^{-3}, and w' and T' are the instantaneous fluctuations of vertical windspeed and temperature about their means. Over a suitable averaging period (e.g., 30 min) this product yields the sensible heat flux, under the assumption of no net transport of dry air, provided the measurements are made at a fast enough sampling rate to encompass all eddy frequencies contributing to transport (Baldocchi et al., 1988; Baker et al., 1992).

MODEL

The SHAW model (Flerchinger and Saxton, 1989) is a one-dimensional simulation that solves coupled differential equations for heat and water flow in soil/residue/snow systems. It is capable of simulating soil freezing and thawing and the accumulation and disappearance of snow, but it does not simulate frost heave. The model requires a number of site parameters, among them soil hydraulic properties (saturated conductivity and moisture characteristic curve), bulk density, and albedo as a function of soil water content. Surface boundary conditions are provided by iteratively solving the surface energy balance at each timestep. The weather data used are those available from standard agricultural weather stations: solar radiation, air temperature, relative humidity, windspeed, and precipitation. Sensible heat flux is calculated with the conventional aerodynamic transport equation,

$$H = \rho_a c_p \frac{T_a - T_s}{r_h} \qquad (12.2)$$

in which T_s and T_a are temperatures at the surface and at screen height, and r_h is the aerodynamic resistance to heat transfer,

$$r_h = \frac{\left(\ln \frac{z}{z_m} + \psi_m\right)\left(\ln \frac{z}{z_h} + \psi_h\right)}{k^2 U} \qquad (12.3)$$

The roughness length for momentum, z_m, is taken as 10 mm in the absence of snow and 1.5 mm when snow is present; z_h is assumed to be equal to 0.2 z_m. The height, z, at which the windspeed, U, is measured, is the screen height minus estimated snow depth. The stability corrections, ψ_m and ψ_h, are calculated using the Monin-Obukhov length, L:

$$L = \frac{-\rho_a c_p T u^{*3}}{kgH} \qquad (12.4)$$

where u^* is friction velocity, k is the von Karman constant (0.4), and g is the acceleration due to gravity. Under stable conditions ($L > 0$), it is assumed that $\psi_m = \psi_h$, and

$$\psi_h = 4.7\frac{z}{L} \qquad\qquad (12.5)$$

When conditions are unstable ($L < 0$), ψ_m is taken as 0.6 ψ_h, and

$$\psi_h = -2\ln\left(\frac{1 + \sqrt{1 - 16\frac{z}{L}}}{2}\right) \qquad\qquad (12.6)$$

We initiated the simulation with measured values for soil water content and temperature profiles obtained on day 330 in 1993, and we ran the model through the winter to day 110 of 1994, using hourly values for air temperature, humidity, windspeed, solar radiation, and precipitation. Model estimates of sensible heat flux and surface temperature were then compared to measured values for the period from day 45 to day 76 in 1994, which encompassed the period of final snowmelt.

RESULTS AND DISCUSSION

The primary comparison of modeled vs. measured sensible heat flux during the 30-day period of the experiment is shown in Figure 12.1A, while Figure 12.1B shows a similar comparison of modeled surface temperature vs. the radiometric surface temperature obtained from an inverted pyrgeometer. Figure 12.1C shows midday albedos for each day as an indicator of the presence of snow, and Figure 12.1D shows measured net radiation. As the period began, the snowpack contained virtually all of the precipitation that had fallen since the beginning of winter in the form of a dense layer of snow and ice, with an approximate thickness of 25 cm and a liquid water equivalent (LWE) of approximately 7 cm. The snowpack melted completely between day 46 and day 51, with a substantial contribution of sensible heat, particularly on day 49. Both the model and the measurements indicate the stabilization of surface temperature at or near 0°C during melt at midday on days 47 and 48 and for a substantial period of time on days 49 to 50. It is noteworthy that the model, when provided with initial conditions from day 330 of 1993 and operated with standard meteorological input data, can accurately predict the occurrence of melting and snowpack disappearance 85 days later, without any interim corrections. Both model and measurements indicate sensible heat flux toward the surface most of the time on days 45 to 49, but the model tends to overestimate the flux, particularly on day 49. The model estimates of surface temperature appear to agree rather well with the measurements during this period, suggesting that the overestimates of the sensible heat flux may be due to errors in parameterization of aerodynamic resistance.

Following the first melt period, the weather turned cold again, but the combination of clear skies and a bare soil surface with low albedo (approximately 0.1) produced midday sensible heat fluxes from surface to atmosphere. Again the model

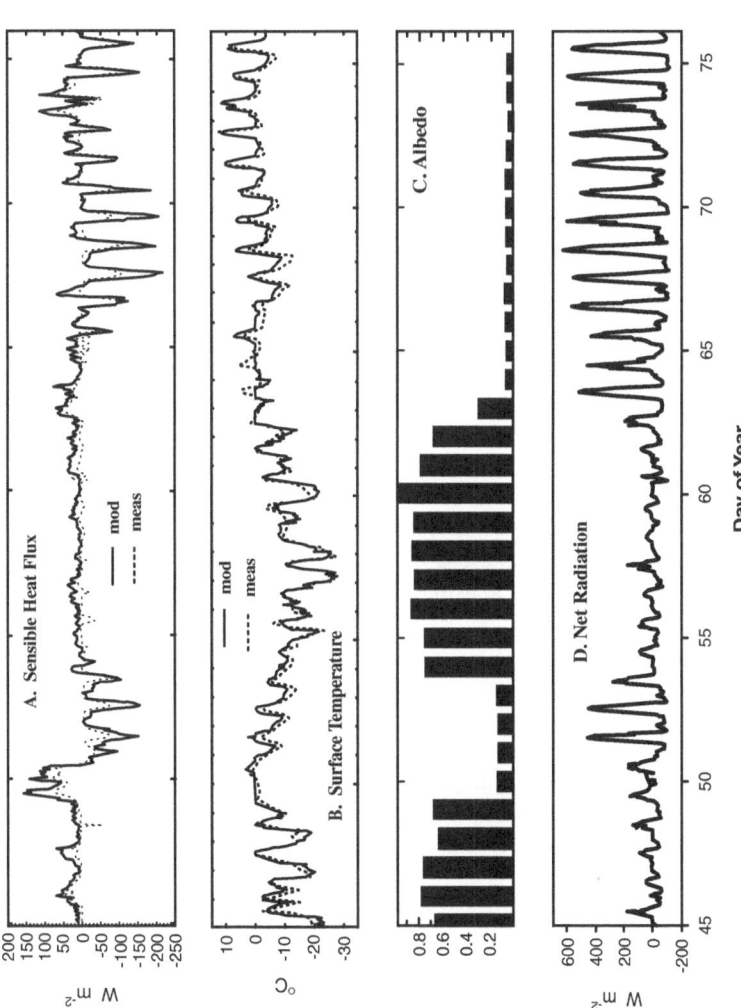

Figure 12.1 Time series for the winter of 1993–1994, Rosemount, Minnesota. (A) Sensible heat flux; (B) Surface temperature; (C) Mean midday albedo — this was calculated from a pair of Kipp & Zonen pyranometers, one facing upward and one inverted; the value of 1.0 on day 60 was due to partial occlusion of the upward-facing pyranometer by morning snowfall, and the albedo of the underlying soil when it is wet is approximately 0.08; (D) Measured net radiation.

and the measurements agree on general trends, but the model tends to overestimate the magnitude of the midday values. In this case the surface temperature comparisons (Figure 12.1B) indicate that the model is overestimating midday surface temperatures. There was new snowfall on days 54 to 56, amounting to approximately 15 cm (1.8 cm LWE). This remained on the ground for several days before the final snowmelt on days 61 to 63. During the snow-covered interval of days 54 to 61, the measured and modeled sensible heat fluxes show little correspondence with one another, but the absolute values of the fluxes are small. The snow was melted by the morning of day 63, but in this instance the model didn't predict disappearance of the snow until afternoon of day 64. As a result, the model and measurements disagree on the direction of the sensible heat flux on days 63 to 64. This is exacerbated by the inherent difficulty of representing a patchy surface in a one-dimensional model. The model retains a reflective surface until the "snow depth" declines to zero, whereas in reality the surface goes through a transition stage that may last for days during which it is an evolving amalgamation of highly reflective, snow-covered patches interspersed with low-albedo, dark, bare soil.

During the final 10 days the soil was bare, so daytime net radiation values were much higher than they had been over the snow. Daytime surface temperatures during this period were consistently warmer than the air, so the direction of sensible heat flux was from the surface toward the atmosphere. The model and the measurements agree on the direction, but the model shows midday overestimates on days 67 to 70. We examined these days in more detail in Figures 12.2A and 12.2B. Figure 12.2A shows the measured and modeled sensible heat fluxes during this time period, and Figure 12.2B shows surface temperatures measured by infrared thermometers and a pyrgeometer, as well as the model estimates. During the daytime on days 67 to 70, when the model overestimates sensible heat flux, the cause appears to be overestimation of surface temperature. On day 71, the model again begins the day with significant overestimates of surface temperature and sensible heat flux, but shortly after noon both of them come into better agreement with measurements. This improvement persists on subsequent days 72 to 74.

The pyrgeometer and the infrared thermometers did not agree with each other, so it is difficult to say *a priori* how large the model errors were, but Figure 12.3 is helpful. It shows logarithmic temperature profiles at midday on four of the days in question, obtained from fine-wire thermocouples. It seems to indicate that the pyrgeometer data may be the most accurate estimator of surface temperature, and it also shows clearly the model overestimates on days 68 and 69. As noted previously, the estimates of both surface temperature and sensible heat flux improve from midday of day 71 on. It is not surprising that surface temperature is most difficult to model in the first few days after the snow disappears. Surface temperature is a controlling variable not only for sensible heat flux, but also for surface evaporation and for soil heat flux, which at this time includes thawing of the soil surface layer. All processes must be correctly simulated in order to arrive at the correct surface temperature. It is particularly difficult to simulate soil heat flux correctly because it depends on thermal conductivity, liquid water content, and ice content of the soil. These all depend on bulk density, which can be quite dynamic during the melt process if any heaving takes place.

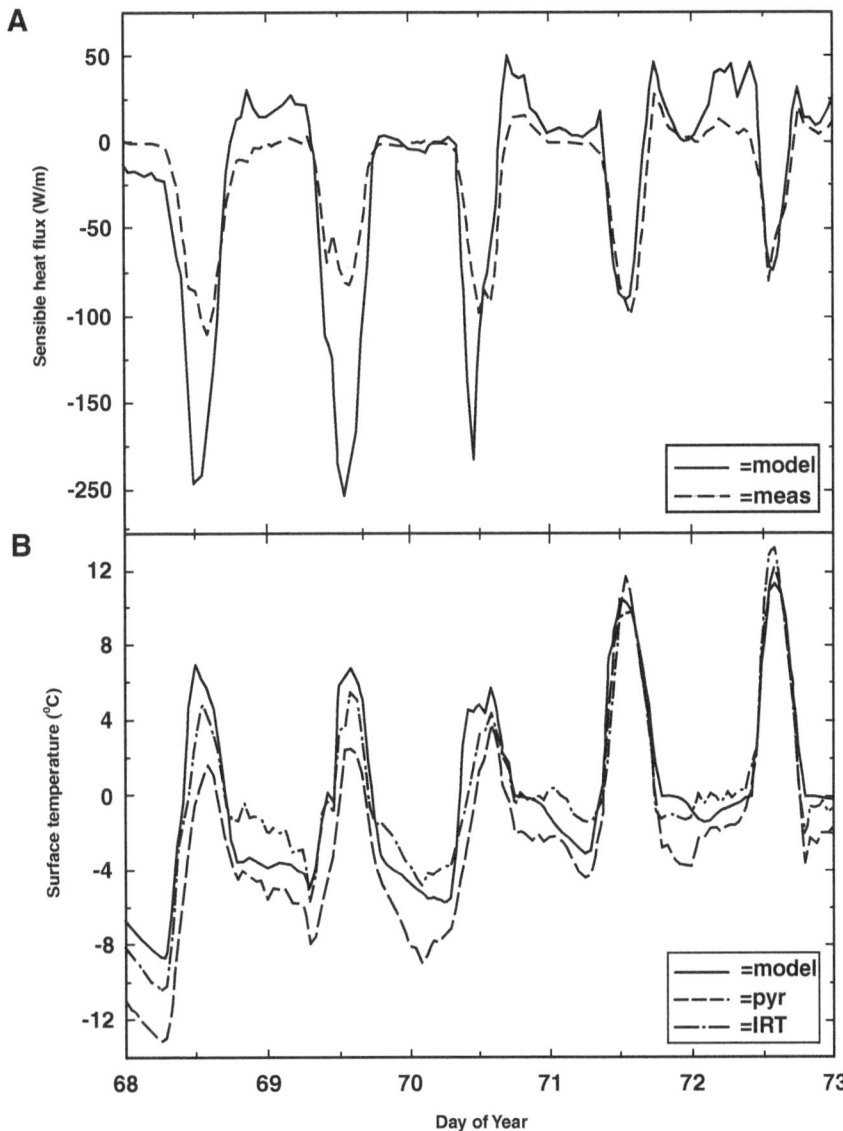

Figure 12.2 (A) Measured and modeled sensible heat flux for days 68–73; (B) measured (both by infrared thermometry and by pyrgeometer) and modeled surface temperature for days 68 to 73.

To assess the accuracy of the model in estimating aerodynamic resistance we first split the data into stable and unstable cases, then considered only those cases where the sensible heat flux was greater than 30 W m^{-2} *and* where the measured and modeled surface–air temperature gradients agreed within 20%. The results are shown in Figure 12.4. The open circles represent unstable conditions, when the heat

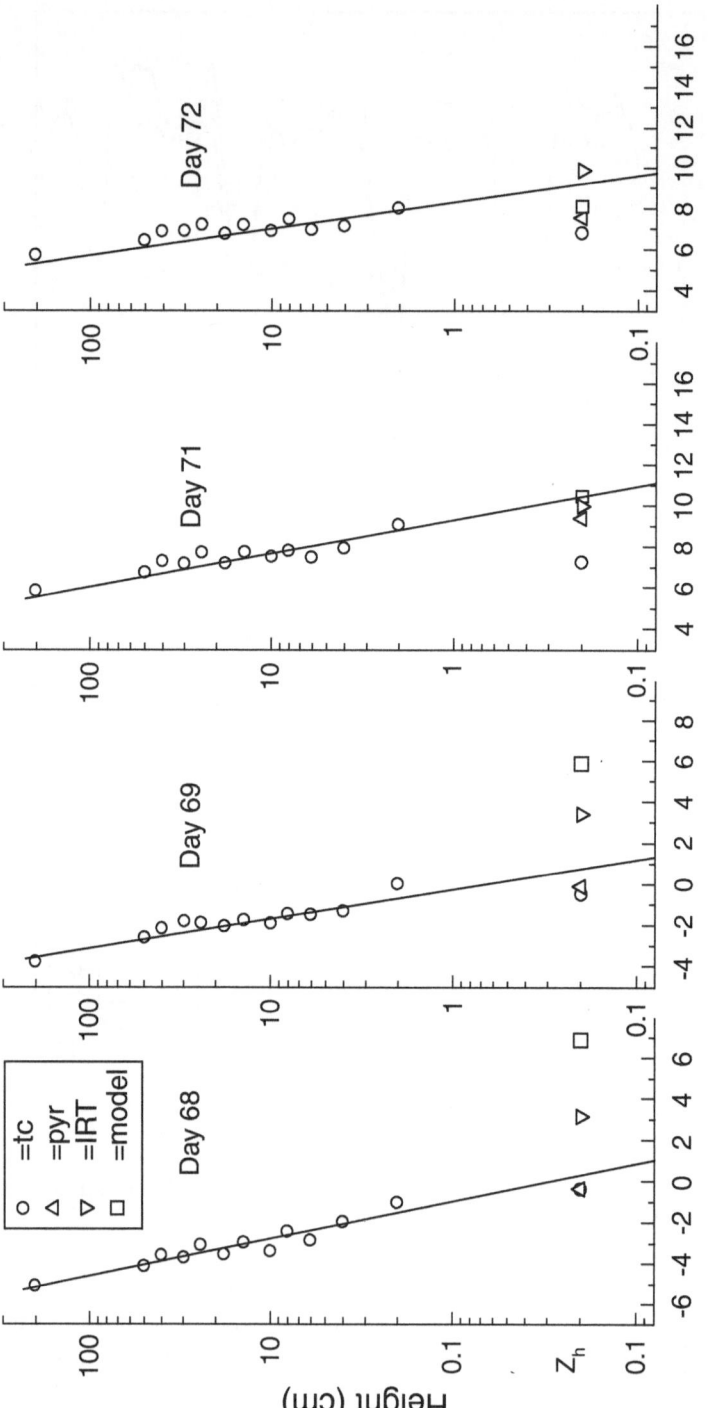

Figure 12.3 Logarithmic profiles of temperature at midday for four selected days, measured with fine-wire thermocouples. For comparisons, the modeled and measured surface temperatures are also shown.

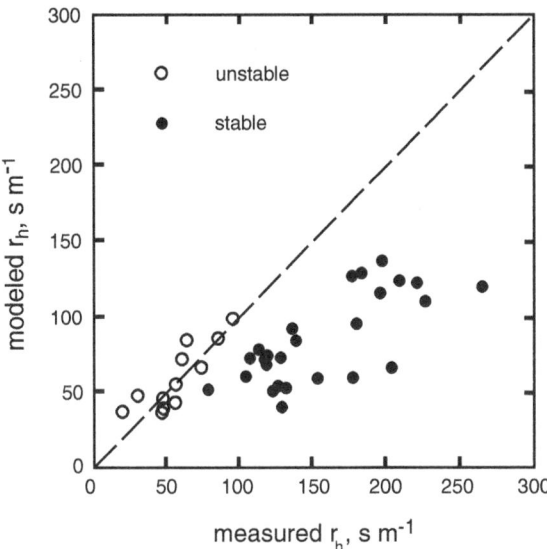

Figure 12.4 Measured and calculated aerodynamic resistances r_h for periods when the absolute value of the sensible heat flux was at least 30 W m^{-2} and the model and measurements of surface temperature agreed within 20%.

flux was from surface to atmosphere, while the closed circles indicate data from stable time periods, when the heat flux was in the opposite direction. The resistances were computed by dividing the measured and simulated sensible heat fluxes by the measured and simulated surface–air temperature gradients, multiplied by the volumetric heat capacity of air. In the unstable case the data are unbiased, showing similar scatter on both sides of the 1:1 line. We tentatively conclude that under unstable conditions the tendency of the model toward overestimation of heat flux shown in Figure 12.2A is due primarily to overestimation of surface temperature, particularly in the first few days after snowmelt.

In the stable case the data are uniformly on one side of the 1:1 line, indicating that the model consistently underestimates aerodynamic resistance during stable conditions. Much of these data were obtained during the snow-covered period. Apparently the parameterization of surface roughness of snow or the stability corrections for stable conditions are incorrect. Moore (1983) summarized a wide range of reported aerodynamic measurements made over snow and concluded that the roughness length for momentum, z_m, varies from a minimum of approximately 1 mm over smooth snow or ice to a maximum of about 5 mm over rough or melting snow. Thus the fixed value of 1.5 mm assumed in the SHAW model is apparently within the range of reported values. Moore further concludes on theoretical grounds that the roughness length for heat, z_h, is at least an order of magnitude smaller than z_m over snow. Kondo and Yamazaki (1990) used a value of 0.13 mm for z_h based on measurements that they had made and considered it to be insensitive to windspeed, but they didn't report any coincident values of z_m. In an extensive theoretical treatment, Andreas (1986) argues that z_h/z_m over snow and ice is not constant, but rather a function of the roughness Reynolds' number, R^* ($= u^* z_m/v$); i.e., it depends on the 'roughness' of the flow regime, decreasing as the flow becomes aerodynamically rougher. For typical u^* and the range of z_m reported by Moore (1983) [which are similar to those measured by Andreas and Claffey (1995) in a more recent study], Andreas' (1986) model produces ratios of 0.01 to 0.1, similar to those of Moore (1983). In SHAW it is taken to be 0.2. If this ratio is too large it could cause underestimation of the aerodynamic resistance to heat transport and hence overestimation of the sensible heat flux. We also cannot exclude the possibility of measurement error. During periods of strong stability the covariance spectrum shifts toward higher frequencies (Baldocchi et al., 1988), with smaller eddies carrying more of the flux. Given the relatively low height of measurement, 2 m, it may be that some of the high frequency components may have been attenuated during these stable periods.

Figures 12.5C and 12.5D show the cumulative sums of sensible heat flux and net radiation over portions of the 31-day period, arbitrarily starting on day 45 and restarting on day 57 following the initial melt and subsequent final snowfall. Both starting points coincide with times at which snow water measurements were made. Figures 12.5A and 12.5B show snow depth and snow water equivalent, the latter expressed in terms of the energy required to melt it. Unfortunately the paucity of snow density measurements limits the extent and strength of interpretations, but it is evident that the initial melt on days 46 to 51 was due almost entirely to sensible heat flux, and the 20 MJ m^{-2} needed to melt the snow nearly matches the cumulative energy measured by the time of disappearance. During the second and final melt period, sensible heat again plays an important role in the initial stages of the melt process, and it is only when the snow is nearly gone that net radiation becomes important. At this point surface heating causes the sensible heat flux to change sign, and by the end of the 31-day period the net contribution of sensible heat to the system is near zero. This is of course misleading, because a casual observer might conclude that radiation drives snowmelt, when in fact radiation does not become important until sensible heat has melted most of the snow. Apparently regional-scale advection of warm air (>0°C) is more important than solar radiation in the initial stages of snowmelt, at least for this location and year. There are occasions where latent heat exchange can also be an important contributor; we have frequently observed condensation on a snowpack from relatively humid air. Unfortunately it is difficult to measure because it is typically accompanied by condensation on the sonic anemometer transducers. This is an example of a situation in which a well-validated model might provide otherwise unobtainable estimates.

CONCLUSIONS

Sensible heat flux is a crucial variable during snowmelt, particularly the early phases of snowmelt prior to the appearance of any underlying soil. Prediction of water release during snowmelt thus requires the ability to accurately model sensible heat flux over snow. In a field test, the SHAW model accurately predicted the date at which snowmelt began after simulating 75 days of heat and mass transport, using

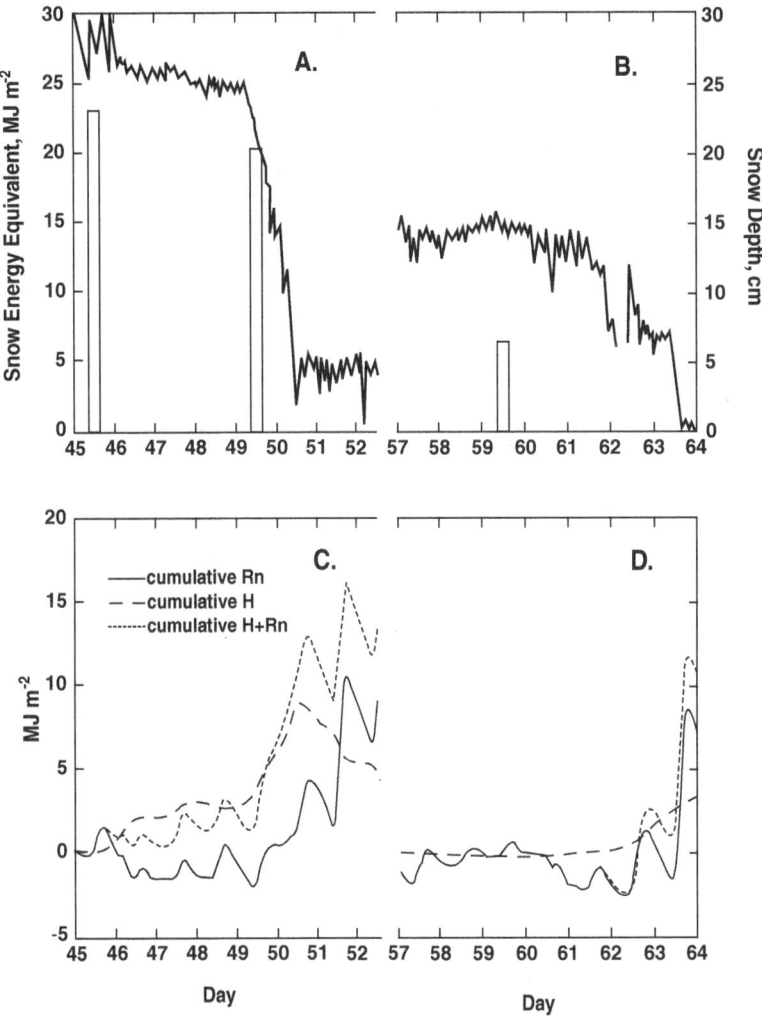

Figure 12.5 Snow depth and snow–water equivalent and cumulative measured sensible heat flux and net radiation. (A) Snow depth (solid line) and snow water equivalent (bars) expressed in terms of the energy required for melting for the first melt period; (B) as in Figure 12.5A, but for the second melt period; (C) cumulative measured sensible heat flux and net radiation during the first melt period; (D) as in Figure 12.5C, but for the second melt period.

only standard meteorological input data. During the subsequent snowmelt and soil thaw sensible heat flux predictions from the SHAW model followed the same temporal trends as the measurements, but tended to overestimate the magnitude of the sensible heat flux, in both directions. During unstable conditions, primarily over bare, thawing soil immediately after disappearance of snow, the errors appear to be due to errors in estimation of surface temperature. During primarily stable conditions when snow is present, the errors appear to be due to the parameterization of surface roughness over snow. These results can be used to modify and improve the model.

REFERENCES

Andreas, E.L. A Theory for the Scalar Roughness and the Scalar Transfer Coefficients over Snow and Sea Ice. *B. Layer Meteorol.* 38, 159–184, 1986.

Andreas, E.L. and K.J. Claffey. Air-Ice Drag Coefficients in the Western Weddell Sea. I. Values Deduced from Profile Measurements. *J. Geophys. Res.* 100, 4821–4831, 1995.

Baker, J.M., J.M. Norman, and W.L. Bland. Field-Scale Application of Flux Measurement by Conditional Sampling. *Agric. Forest Meteorol.* 62, 31–52, 1992.

Baldocchi, D.D., B.B. Hicks, and T.P. Meyers. Measuring Biosphere-Atmosphere Exchanges of Biologically Related Gases with Micrometeorological Methods. *Ecology* 69, 1331–1340, 1988.

Flerchinger, G.N. and K.E. Saxton. Simultaneous Heat and Water Model of a Freezing Snow-Residue-Soil System. I. Theory and Development. *Trans. Am. Soc. Agric. Eng.* 32(2), 565–571, 1989.

Kondo, J. and T. Yamazaki. A Prediction Model for Snowmelt, Snow Surface Temperature and Freezing Depth Using a Heat Balance Method. *J. Appl. Meteorol.* 29, 375–384, 1990.

Kuehnast, E.L., D.G. Baker, and J.L. Zandlo. Climate of Minnesota. XII. Duration and Depth of Snow Cover. Tech. Bull. 333, University of Minnesota Agric. Exp. Stn., St. Paul, 1982.

Moore, R.D. On the Use of Bulk Aerodynamic Formulae over Melting Snow. *Nordic Hydrol.*, 1983, 193–206, 1983.

CHAPTER 13

Retention Kinetics of Heavy Metals in Soils: Modeling Approaches

H. M. Selim and I. K. Iskandar

CONTENTS

INTRODUCTION

The problem of identifying the fate of heavy metals in soils must account for retention reactions and transport of the various species in the soil environment. In fact, heavy metals in soils can be involved in a series of complex chemical and biological interactions. Oxidation–reduction, precipitation and dissolution, volatilization, and surface and solution phase complexation are among these reactions. A number of scientists investigated soil properties that significantly affect the behavior of heavy metals such as copper in soils (e.g., Aringhieri et al., 1985; Buchter et al., 1989). Barrow (1989) showed that the use of single-reaction models is not adequate, since such models describe the fate of individual species with no consideration to the simultaneous interactions with others in the soil system. The work of Amacher et al. (1986) showed that sorption–desorption of several heavy metals from batch studies on several soils was not adequately described by use of single reactions involving equilibrium Langmuir or Freundlich isotherms. They also found that a first-order kinetic reaction was not capable of describing changes in Cd, Cr, and Hg concentrations in soil solution with time. Aringhieri et al. (1985) showed that retention of Cd and Cu by an organic soil was strongly time dependent. Cernik et al. (1994) described Cu and Zn distributions with depth in a contaminated soil near a metal smelter, and a modified (equilibrium) Freundlich sorption model was incorporated into their convective–dispersion equation. Recently, Cu leaching in soil columns was described based on equilibrium adsorption–desorption, coupled with kinetic solubilization (Montero et al., 1994).

A disadvantage of several empirical approaches lies in the basic assumption of local equilibrium of the governing reactions. Alternatives to equilibrium-based approaches are the multisite or multireaction models that deal with the multiple interactions of a single solute species in the soil environment. Multiple interaction approaches commonly assume that a fraction of the total sites is time dependent (kinetic in nature), whereas the remaining fraction interacts rapidly or instantaneously with that in the soil solution. Nonlinear equilibrium (Freundlich) and first- or nth-order kinetic reactions are the associated processes (for a review see Selim, 1992). Another two-site approach was proposed by Theis et al. (1988) for Cd mobility and adsorption on goethite. The nature of reactions for both sites was assumed to be governed by second-order kinetic reactions. Reactions were assumed to be consecutive where the second reaction was irreversible in nature. Amacher et al. (1988) developed a multireaction model that includes concurrent and concurrent-consecutive processes of the nonlinear kinetic type. The model was capable of describing the retention behavior of Cd and Cr(VI) with time for several soils. Selim et al. (1992) developed a multicomponent approach that accounts for an equilibrium ion exchange and a specific sorption process based on a second-order (Langmuir) kinetic reaction. The multicomponent model adequately predicted the observed breakthrough results where a pronounced snowplow or chromatographic effect was observed. Effluent peak concentrations were three- to fivefold that of the input Cd pulse. This snowplow effect has been observed by Starr and Parlange (1979) and is a result of the release of sorbed species from matrix surfaces in response to a large change in total concentration (or ionic strength) of the applied (input) solutions.

This chapter emphasizes major features of retention reactions and modeling approaches of heavy metals in the soil environment. Single reaction models of the reversible and irreversible kinetic type are discussed first. Models of the multiple-reaction type, including the two-site equilibrium-kinetic models, the concurrent and consecutive multireaction models, and the second-order approach, are also discussed.

SINGLE RETENTION APPROACHES

Freundlich and Linear

For several heavy metals (e.g., Cr, Cu, Zn, Cd, and Hg), retention and release reactions in the soil solution have been observed to be strongly time-dependent. Recent studies on the kinetic behavior of the fate of several heavy metals include Harter (1984), Aringhieri et al. (1985), and Amacher et al. (1986) among others. A number of empirical models have been proposed to describe kinetic retention and release reactions of solutes in the solution phase. The first-order kinetic approach is perhaps the earliest single reaction form used to describe time-dependent sorption, which may be expressed as

$$\rho\frac{\partial S}{\partial t} = k_f \Theta C - k_b \rho S \qquad (13.1)$$

where C is the solute concentration in solution (μg/mL), S is the amount of solute sorbed or retained by the soil (μg/g soil), ρ is the soil bulk density (g/cm^3), Θ is the volumetric soil–water content (cm^3/cm^3), and t is time (h). The reaction described in Equation 13.1 is fully reversible between the solute species present in solution and that sorbed by soil matrix surfaces, where k_f and k_b represent the forward and backward rate coefficients (h^{-1}). The first-order reaction was first incorporated into the classical (convective–dispersive) transport equation by Lapidus and Amundson (1952) to describe solute retention during transport under steady water flow conditions.

Integration of the first-order reaction (Equation 13.1) subject to initial conditions of $C = C_i$ and $S = 0$ at $t = 0$ yields a system of linear sorption isotherms. That is, for any reaction time t, a linear relation between S and C is obtained. However, linear isotherms are not often encountered except for selected cations and heavy metals at low concentrations. In contrast, nonlinear retention behavior is commonly observed for several solutes, as depicted by the nonlinear isotherms for Cu for Cecil and McLaren soils shown in Figure 13.1. As a result, the single reaction given by Equation 13.1 has been extended to include the nonlinear kinetic form,

$$\rho\frac{\partial S}{\partial t} = k_f \Theta C^m - k_b \rho S \qquad (13.2)$$

where m is a dimensionless parameter commonly less than unity and represents the order of the nonlinear reaction. For both single kinetic forms (Equations 13.1 and 13.2), the magnitude of the rate coefficients dictates the extent of the kinetic behavior of the reaction. For small values of k_f and k_b, the rate of retention is slow, and strong kinetic dependence is anticipated. In contrast, for large values of k_f and k_b, the retention reaction is a rapid one and should approach quasi-equilibrium in a relatively short time. In fact, at large times ($t \rightarrow \infty$) equilibrium is attained and the rate of

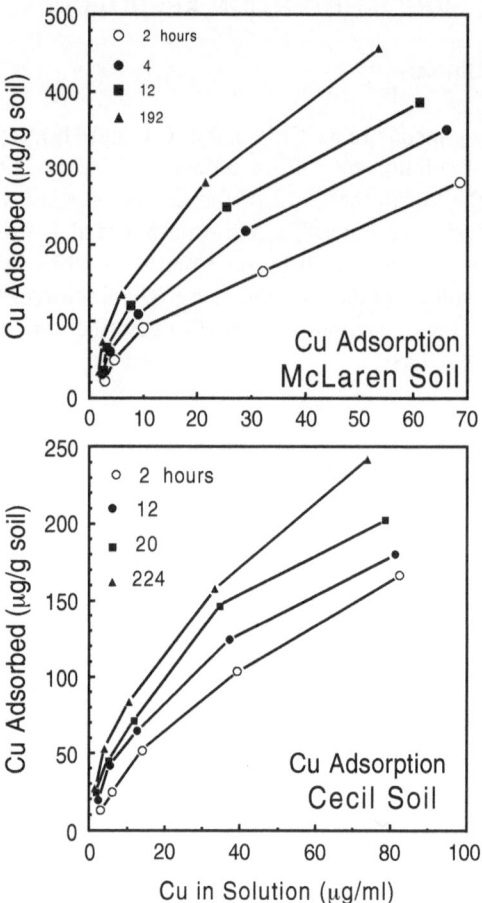

Figure 13.1 Cu adsorption isotherms at several reaction times in Cecil and McLaren soils.

retention ($\partial S/\partial t$) approaches zero, and the above equation yields a form analogous to the Freundlich equilibrium equation:

$$S = K_d C^m, \qquad \text{where} \qquad K_d = \left(\frac{\Theta}{\rho}\right)\frac{k_f}{k_b} \qquad (13.3)$$

where K_d is the distribution coefficient (cm³/g). Therefore, for linear or Freundlich isotherms, one may regard K_d as the ratio of the rate coefficient for sorption (forward reaction) to that for desorption or release (backward reaction).

Langmuir

The Langmuir isotherm is most commonly encountered in soils. It was developed to describe the adsorption of gases by solids where a finite number of adsorption sites on the surface is assumed (Langmuir, 1918). As a result, a major advantage of the Langmuir equation over linear and Freundlich types is that a maximum sorption capacity is incorporated into the formulation of the model, which may be regarded as a measure of the number of available retention sites on the solid phase. The standard form of the Langmuir equation is

$$\frac{S}{S_{max}} = \frac{\omega C}{1 + \omega C}$$
(13.4)

where ω and S_{max} are adjustable parameters. Here ω (mL μg^{-1}) is a measure of the bond strength of molecules on the matrix surface, and S_{max} ($\mu g\ g^{-1}$ of soil) is the maximum sorption capacity or total amount of available sites per unit soil mass. It was recognized that the Langmuir isotherm is the most commonly used (Sposito, 1984) and is referred to as the L-curve isotherm. Moreover, Langmuir isotherms were used successfully to describe Cd, Cu, Pb, and Zn retention in soils. Figure 13.2 shows experimental and fitted isotherm examples of the use of the Langmuir equation to describe Cu retention in Cecil and McLaren soils after 192 h of reaction.

Hysteresis

Adsorption–desorption hysteretic behavior has been observed by several scientists. Examples of hysteretic behavior for Cu adsorption–desorption for a McLaren soil is shown in Figure 13.3. Cu desorption shows significant hysteresis or nonsingularity behavior, which was apparent at high initial concentrations. Based on the hysteresis behavior for several initial concentrations (results not shown), desorption results suggest that part of the adsorbed Cu was either not

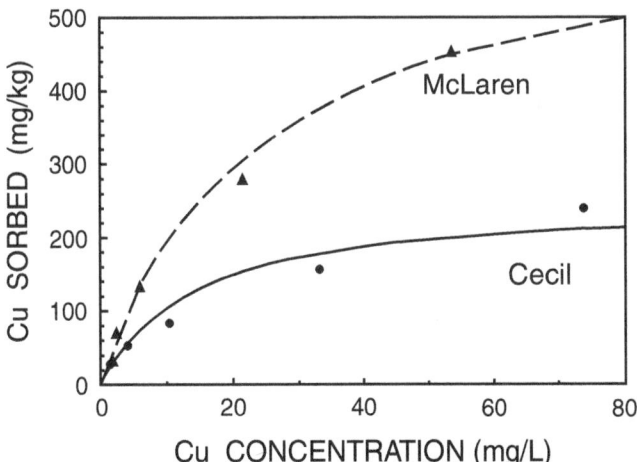

Figure 13.2 Retention isotherms for Cu after 8 days of reactions for Cecil and McLaren soils. Solid curves are calculated isotherms using equilibrium Langmuir model.

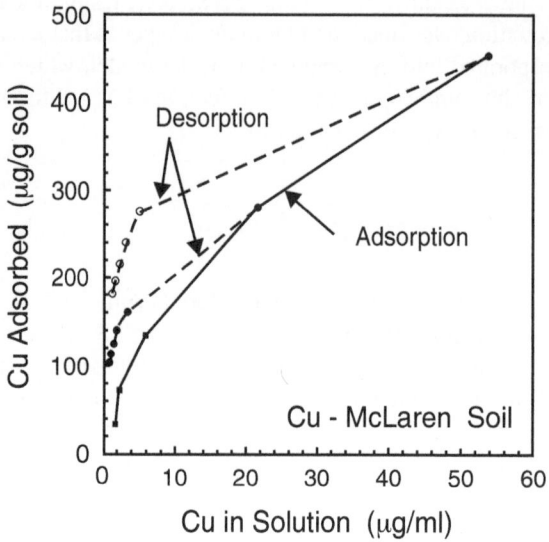

Figure 13.3 Cu adsorption and desorption isotherms for McLaren soil.

easily desorbed or becomes nondesorbable by forming a strong interaction with the soil matrix. Selim et al. (1976) showed that nonsingularity or hysteresis may partly result from failure to achieve equilibrium during adsorption and desorption. If adsorption as well as desorption were carried out for times sufficient for equilibrium to be attained, or if the kinetic rate coefficients were sufficiently large, such hysteretic behavior would be minimized.

Irreversible Reactions

Irreversible processes of various solute species present in the soil environment account for various (sink/source) reactions, including precipitation/dissolutions, mineralization, immobilization, biological transformations, volatilization, and radioactive decay, among others. First-order kinetic reactions have been used to quantify the irreversible retention processes by several authors. Models that account for first-order kinetic and sequential first-order (irreversible) decay reactions include those of Cho (1971), Selim and Iskandar (1981), Rasmuson and Neretienks (1981), and Amacher et al. (1988). The form for first-order irreversible retention rate is

$$Q = k_{irr}\Theta C \qquad (13.5)$$

where k_{irr} is the irreversible rate of reaction (h^{-1}). This mechanism has been used to describe various N transformation processes, P and heavy metal adsorption/precipitation, and radionuclide decay. Description of precipitation reactions that involve secondary nucleation is not an easy task, and it is often difficult to distinguish between precipitation and adsorption.

Specific Sorption

Specific sorption on high affinity sites has been considered as an irreversible process. A second-order approach was modified to describe irreversible or weakly reversible retention when the rate of desorption or release is small, i.e., when the backward rate coefficient k_b approaches zero (see Selim et al., 1992). Therefore, a second-order irreversible process can be expressed as

$$\rho \frac{\partial S}{\partial t} = k_f \Theta (S_T - S) C \qquad (13.6)$$

where only two parameters, S_T and k_f, are required to account for irreversible retention. The term S_T represents the total amount of specific sites ($\mu g \ g^{-1}$ of soil). For several metal ions, including Cd, Ni, Co, and Zn, specific sorption has been shown to be time dependent. Therefore, the use of a kinetic rather than an equilibrium sorption mechanism is recommended. A major advantage of the formulation of irreversible reaction Equation 13.7 is that a sorption maximum is achieved when all unfilled sites become occupied (i.e., $S \rightarrow S_T$). In contrast, for the first-order type Equation 13.6, an irreversible sorption maximum is not attained as metal ion concentration increases.

MULTIPLE RETENTION APPROACHES

A number of investigations indicated that a single reaction of the equilibrium or kinetic type did not adequately describe the retention (adsorption–desorption) of dissolved chemicals in several soils. The inadequacy of single reactions is not surprising, since they describe the behavior of only one species with no consideration to accompanying simultaneous reactions of other species in the soil system. Multicomponent models consider a number of processes governing several species in the soil solution including ion exchange, complexation, dissolution/precipitation, and competitive adsorption. However, multicomponent models often consider the local equilibrium assumption (LEA) to be valid. On the other hand, multisite or multireaction models deal with the multiple interactions of one species in the soil environment. Such models are empirical in nature and are based on the assumption that a fraction of the total sites are highly kinetic, whereas the remaining fraction of sites interact slowly or instantaneously with that in the soil solution (Selim et al., 1976; Jardine et al., 1985). Nonlinear equilibrium (Freundlich) and first- or nth-order kinetic reactions were the associated processes.

The two-site approach proved successful in describing observed breakthrough results and has been used by several scientists including Jardine et al. (1985) and Parker and Jardine (1986), among others. The model proved successful in describing the retention and transport of several dissolved substances including aluminum, phosphorus, potassium, cadmium, chromium, and methyl bromide. However, there are several inherent disadvantages of the two-site model. First, the reaction mechanisms are restricted to those that are fully reversible. Moreover, the model does not account for possible consecutive-type solute interactions in the soil system. Another two-site approach was proposed by Theis et al. (1988) for Cd mobility and adsorption on goethite. They assumed the nature of reactions for both sites to be governed by second-order kinetic reactions. The reactions were assumed to be consecutive where the second reaction was irreversible in nature. Other empirical approaches include the consecutive (equilibrium-kinetic) model of Barrow and Shaw (1979). Here an adsorbed (surface) phase was assumed to be in direct equilibrium with that in the solution phase and slowly (and reversibly) with an absorbed (internal) phase within the soil matrix. Another approach is that of van der Zee and van Riemsdijk (1986), in which a reversible reaction for P sorption–desorption was governed according to the Langmuir kinetic model (Equation 13.4). In addition, precipitation reaction was accounted for as a (kinetic) diffusion process of P through a thin layer of metal phosphate coating that surrounds metal oxides. Metal oxide is assumed to be con-verted to metal phosphate in the reaction zone by a precipitation-like reaction.

Amacher et al. (1988) and Selim et al. (1989) proposed a simplified model that accounts for multiple reactions of solutes during transport in soils. In addition to the soil solution phase (C) of a solute in the soil, four other phases representing solute retained by the soil matrix (S_e, S_1, S_2, S_3, and S_{irr}) were also considered. A schematic of the multireaction model is shown in Figure 13.4. We assume Se to be governed by an equilibrium Freundlich reaction, whereas S_1 and S_2 were governed by nonlinear kinetic reactions,

$$S_e = K_d C^b \tag{13.7}$$

$$\frac{\partial S_1}{\partial t} = k_1 \frac{\Theta}{\rho} C^n - k_2 S_1 \tag{13.8}$$

$$\rho \frac{\partial_2}{\partial} = k_3 \Theta C^m - (k_4 + k_5)\rho_2 + k_6 \rho_3 \tag{13.9}$$

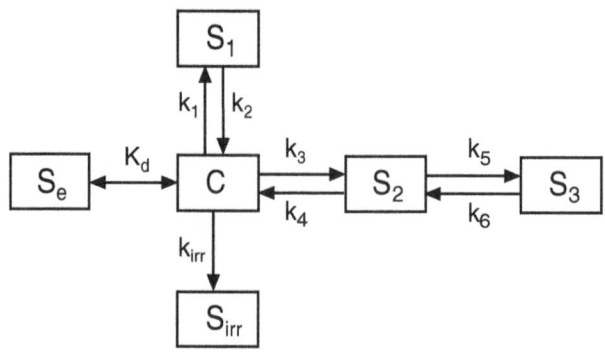

Figure 13.4 Schematic diagram of the multireaction model.

where k_1 to k_6 are the associated rates coefficients (h^{-1}). These two phases (S_1 and S_2) may be regarded as the amounts sorbed on surfaces of soil particles and chemically bound to Al and Fe oxide surfaces or other types of surfaces.

Amacher et al. (1988) pointed out that it is not necessary to have a prior knowledge of the exact retention mechanisms for these reactions to be applicable. Moreover, these phases may be characterized by kinetic sorption and release behavior to the soil solution, and thus they are susceptible to leaching in the soil. In addition, the primary difference between these two phases not only lies in the difference in their kinetic behavior, but also in the degree of nonlinearity as indicated by the parameters n and m. The consecutive reaction between S_2 and S_3 represents slow reaction as a result of further rearrangements of solute retained on matrix surfaces. Incorporation of S_3 in the model permits the description of the frequently observed very slow release of solute from the soil. As a result, this strongly retained phase was represented by

$$\frac{\partial S_3}{\partial t} = k_5 S_2 - k_6 S_3 \qquad (13.10)$$

The multireaction model also considers irreversible solute removal via a retention sink term Q to account for irreversible reactions such as precipitation dissolution, mineralization, and immobilization, among others. We expressed the sink term as a first-order kinetic process in a similar fashion to that given by Equation 13.6, where k_{irr} is the associated rate coefficient (h^{-1}).

The multireaction model given by Equations 13.6 to 13.10 was incorporated into the classical convective–dispersion transport equation (CDE), which can be expressed as (Selim et al., 1989)

$$\Theta R \frac{\partial C}{\partial t} + \rho \left(\frac{\partial S_1}{\partial t} + \frac{\partial S_2}{\partial t} \right) = \Theta D \frac{\partial^2 C}{\partial z^2} - v \frac{\partial C}{\partial z} - Q \qquad (13.11)$$

where D is the hydrodynamic dispersion coefficient (cm^2 h^{-1}), v is Darcy's water flux density (cm h^{-1}), and z is soil depth (cm). The term R is dimensionless and referred to here as the Freundlich retardation coefficient that is concentration (C) dependent,

$$R = 1 + \left(\frac{\rho}{\theta} \right) K_d b C^{b-1} \qquad (13.12)$$

For the linear case where the exponent b in Equation 13.7 is unity, the well-known retardation factor R is obtained:

$$R = 1 + \left(\frac{\rho}{\theta} \right) K_d \qquad (13.13)$$

which is constant. The multireaction model was capable of describing the kinetic retention behavior of P, Cd, Cr, Zn, and Hg based on batch data sets for several soils. For example, the solid lines shown in Figure 13.5 are multireaction model predictions of Cd vs. time for the same initial concentration (C_o = 1 µg/mL) for all five soils (Selim, 1989). To test how well this model is capable of predicting the kinetics of solute retention patterns for several species, Amacher et al. (1988) examined several model variations ranging from those where all reactions were included to model variations where only three phases (C, S_1, and S_{irr}) were considered. A number of model variations were capable of producing indistinguishable simulations of the data. Such a conclusion was based on retention data sets from batch studies for Cd, Cr(VI), and Hg for several soils. A similar conclusion was made by other investigators. For example, it was not possible to determine whether the irreversible reaction is concurrent or consecutive, since both variations provided similar overall fit to the batch data.

ION EXCHANGE RETENTION

The ion exchange retention mechanism is commonly considered as a rapid reaction involving the nonspecific sorption type. The mechanism is a fully reversible reaction between ions in the soil solution and those retained on charged matrix

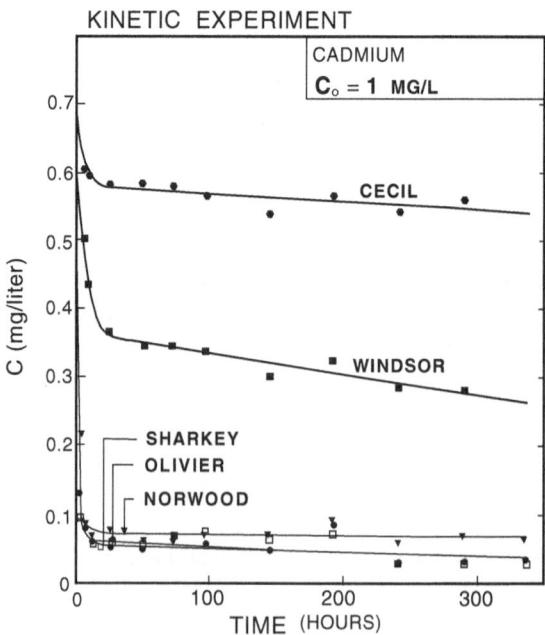

Figure 13.5 Cadmium (Cd) concentration vs. time of reaction for five soils with initial Cd concentration (C_o) = 1 µg/mL.

surfaces. The exchange reaction for two competing ions i and j, having valences v_i and v_j, respectively, may be written as

$$^T K_{ij} = \frac{(a_i^* / a_i)^{v_j}}{(a_j^* / a_j)^{v_i}}$$

(13.14)

where $^T K_{ij}$ denotes the thermodynamic equilibrium constant and a and a^* (omitting the subscripts) are the ion activities in soil solution and on the exchanger surfaces, respectively. For the case of binary homovalent ions, a generic selectivity coefficient K_{ij} (Rubin and James, 1973) or a separation factor for the affinity of ions on exchange sites is often used. Examples of calculated and measured homovalent ion exchange isotherms are illustrated in Figure 13.6 for Cd–Ca for two soils (Selim et al., 1992). For $K_{CdCa} > 1$, sorption of ion Cd is preferred and the isotherms are convex, whereas for $K_{CdCa} < 1$, sorption affinity is apposite and the isotherms are concave. The assumption of equilibrium ion exchange reaction has been employed to describe sorption of heavy metals in soils by several investigators (Abd-Elfattah and Wada, 1981; Harmsen, 1977; Bittel and Miller, 1974; Selim et al., 1992; Hinz and Selim, 1994). In general, the affinity of heavy metals increases with decreasing heavy metal fraction on exchanger surfaces. Using an empirical selectivity coefficient, Abd-Elfattah and Wada (1981) showed that Zn affinity increased up to two orders of magnitude for low Zn surface coverage in a Ca background solution. The Rothmund–Kornfeld approach incorporates variable selectivity based on the amount adsorbed (s_i) or exchanger composition. The approach is empirical and provides a simple equation that incorporated

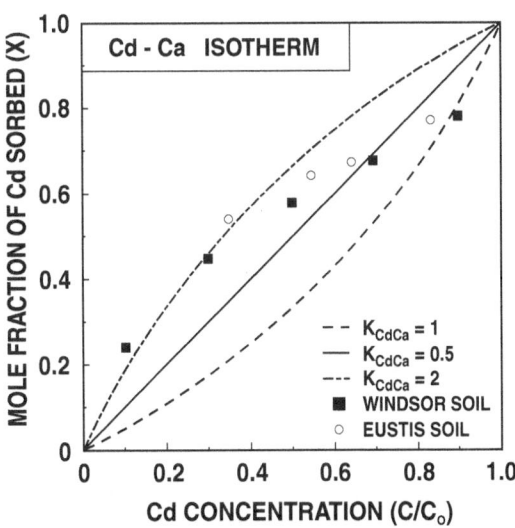

Figure 13.6 Cadmium–calcium exchange isotherm for Windsor and Eustis soils. Solid and dashed curves are simulations using different selectivities (K_{CdCa}).

the characteristic shape of binary exchange isotherms as a function of s_i as well as the total solution concentration in solution (C_T). Harmsen (1977) and Bond and Phillips (1990) expressed the Rothmund–Kornfeld equation as

$$\frac{(s_i)^{v_j}}{(s_j)^{v_i}} = {}^R K_{ij} \left[\frac{(c_i)^{v_j}}{(c_j)^{v_i}} \right]^n \tag{13.15}$$

where n is a dimensionless empirical parameter associated with the ion pair i-j and ${}^R K_{ij}$ is the Rothmund–Kornfeld selectivity coefficient. The above equation is best known as a simple form of the Freundlich equation that applies to ion exchange processes. As pointed out by Harmsen (1977), the Freundlich equation may be considered as an approximation of the Rothmund-Kornfeld equation valid for $s_i \ll s_j$ and $c_i \ll c_j$, where

$$s_i = {}^R K_{ij} (c_i)^n \tag{13.16}$$

The ion exchange isotherms in Figure 13.7 show the relative amount of Zn and Cd adsorbed as a function of relative solution concentration along with best-fit isotherms based on the Rothmund–Kornfeld equation for two acidic soils (Hinz and Selim, 1994). The diagonal line represents a nonpreference isotherm (${}^R K_{ij} = 1$, n = 1), where competing ions (Ca-Zn or Ca-Cd) have equal affinity for exchange sites. The sigmoidal shapes of the isotherms reveal that Zn and Cd sorption exhibit high affinity at low concentrations, whereas Ca exhibits high affinity at high heavy metal concentrations. Examples of transport behavior of Zn and Cd when variable ionic strength (or total concentration) prevailed in the soil columns are presented in Figures 13.8 and 13.9, respectively (Hinz and Selim, 1994). Since the total concentration of the Zn and Cd input pulse solutions were much lower than that of the displacing Ca solution, chromatographic peaks were observed. Early appearance of Zn was well described by the predicted BTC (dashed curves) where equal Ca-Zn exchange affinity was assumed. In fact, the chromatographic effect for Ca and Zn was adequately described by the equal affinity BTCs. However, the Rothmund–Kornfeld equation predictions (solid curves) for both data sets were not satisfactory. These predictions are indicative of strong Zn and Cd affinity (compared to Ca) at low Zn and Cd concentrations based on parameter estimates of ion exchange isotherms using the Rothmund–Kornfeld approach.

Kinetic Ion Exchange

An extensive list of cations (and anions) that exhibited kinetic ion exchange behavior in soils was compiled by Sparks (1989), e.g., Al, NH_4, K, and several heavy metal cations. According to Ogwada and Sparks (1986), kinetic ion exchange behavior was probably due to mass transfer (or diffusion) and chemical kinetic processes. The proposed approach was analogous to mass transfer or diffusion between the solid and solution phase such that, for ion species i,

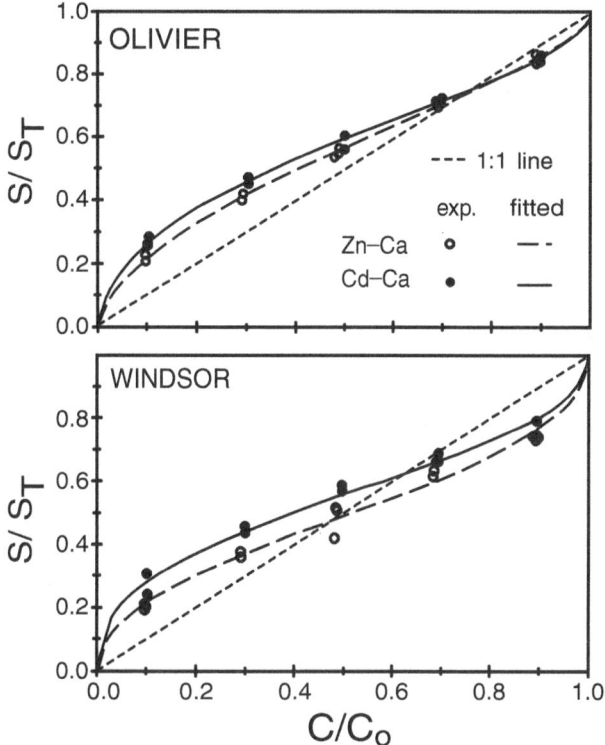

Figure 13.7 Ion exchange isotherms of Cd-Ca and Zn-Ca for Olivier and Windsor soils (relative concentration (C/C_o) vs. the sorbed fraction (S/ST). Solid and dashed curves are fitted using the Rothmund–Kornfeld equation.

$$\frac{\partial s_i}{\partial t} = \alpha\left(s_i^* - s_i\right) \qquad (13.17)$$

where at any time t, the symbol s_i denotes the amount sorbed where s_i^* is the amount sorbed at equilibrium, and α is an apparent rate coefficient (d^{-1}) for the kinetic-type sites. In Equation 13.17, the amount sorbed at equilibrium s_i^* is calculated using the respective isowtherm relations similar to Equation 13.14. Expressions similar to Equation 13.17 have been used to describe mass transfer between mobile and immobile water as well as chemical kinetics (Parker and Jardine, 1986). For large α, s_i approaches s_i^* in a relatively short time and equilibrium is rapidly achieved, whereas for small α, kinetic behavior should be dominant for extended periods of time. To test the capability of the competitive model, two data sets from multiple pulse applications are illustrated. Figures 13.10 and 13.11 are for Windsor soil where Cd pulse applications were 10 and 100 mg L^{-1}, respectively (Selim et al., 1992). For all multiple pulses, the ion exchange

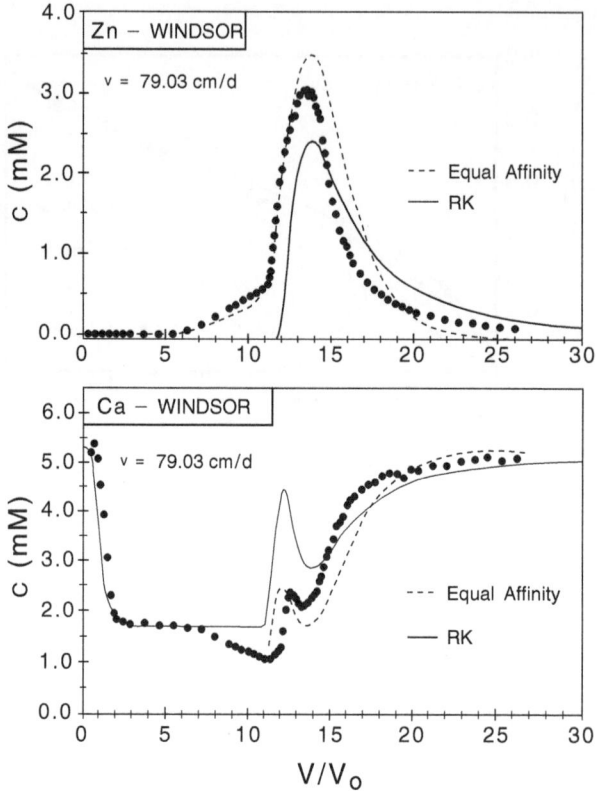

Figure 13.8 Zn and Ca breakthrough curves in Windsor soil column at variable ionic strength. Predictions were based on equal affinity ($K_1 = 1$) and the Rothmund–Kornfeld (RK) equation.

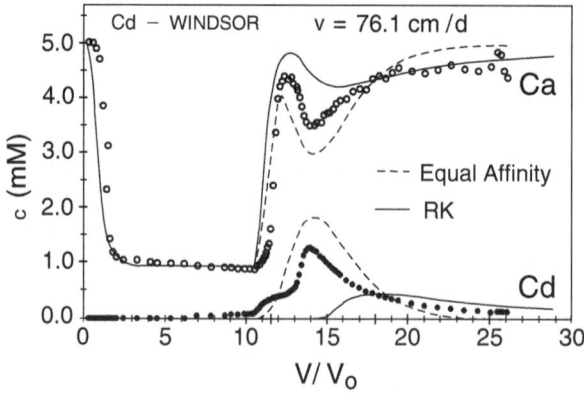

Figure 13.9 Cd and Ca breakthrough curves in a Windsor soil column at variable ionic strength. Predictions were based on equal affinity ($K_{12} = 1$) and the Rothmund-Kornfeld approach.

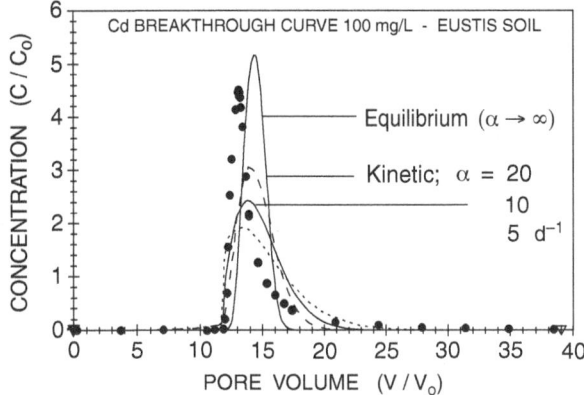

Figure 13.10 Measured (closed circles) and predicted breakthrough curves in Windsor soil column for three Cd pulses of C_o = 10 mg L^{-1}. Curves are predictions using equilibrium and kinetic ion exchange with different α's.

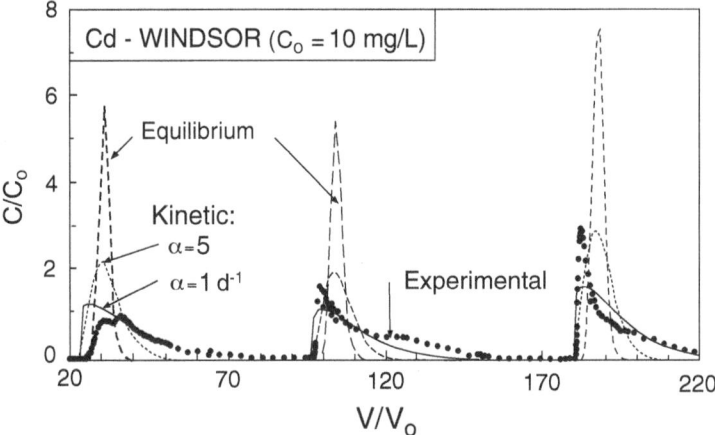

Figure 13.11 Measured (closed circles) and predicted breakthrough curves in Windsor soil column for three Cd pulses of C_o = 100 mg L^{-1}. Curves are predictions using equilibrium and kinetic ion exchange with different α's.

approach well predicted the position of the BTC peaks. In fact, the assumption of equilibrium ion exchange adequately predicted the observed snowplow effect for the two Windsor data sets. The observed slow release or extensive tailing of Cd in the effluent is perhaps an indication of highly kinetic reversible reactions as well as the slow release of Cd retained strongly on specific sorption sites. As a result, the use of kinetic ion exchange provided better description of the multiple pulses for both columns. The value for the transfer coefficient α used with the competitive model varied from 1 to 5 d^{-1} for both soils.

FACTORS AFFECTING RATE OF REACTIONS

Amacher et al. (1988) found that the magnitude of the rate coefficients (k_1 to k_6, k_{irr}) of the multireaction model that provided the predictions shown were highly dependent on the initial (input) concentration C_o. This was an indication that although the model is successful in describing kinetic data for a given C_o, the same rate coefficients cannot be used to describe data for substantially different initial concentrations. Thus, this multireaction model may be considered as an oversimplification that does not provide a complete description of the actual processes that occur during sorption/desorption of solutes in soils and is best considered as a representation of an apparent rate law, rather than a mechanistic rate law. In fact, according to Amacher (1991), a kinetic study is not complete unless the effects of various experimental variables on the experimental rate functions are determined. In addition to concentrations (both solute and solid phase), temperature, pH, ionic strength, and solution composition are known to affect the rate of reactions.

In the literature, the effect of temperature on rate of reactions is one of the most commonly reported. As the temperature increases, the rate of reactions usually increases. The dependence of the rate of reaction on temperature follows the well-known Arrhenius equation:

$$k = Ae^{-E/RT} \tag{13.18}$$

where k is the rate coefficient for a given reaction, A is a frequency factor and has the units of k, E is the activation energy of reaction (kJ mol^{-1}), R is the universal gas constant (J K^{-1} mol^{-1}), and T is absolute temperature in kelvins. For reversible reactions between a solute species in solution C and that sorbed S,

$$C \underset{k_b}{\overset{k_f}{\rightleftarrows}} S \tag{13.19}$$

the forward (k_f) and backward (k_b) rate coefficients are directly dependent on the activation energies. For the simple case of a first-order reaction (see Equation 13.1), the thermodynamic equilibrium constant (K) is the ratio of the forward and backward rate coefficients ($K = k_f/k_b$). Thus, when $K = 1$, activation energies in the forward and reverse directions are equal and a solute has an equal probability of being in the soil solution phase (C) or the solid phase (S). For $K < 1$, the forward activation energy is less than the reverse.

Ogwada and Sparks (1986a) have shown that thermodynamic parameters calculated from rate coefficients for diffusion-controlled exchange reactions compare well in trend, and these parameters lead to the same conclusions as those calculated from equilibrium data or kinetic data where diffusion is not rate limiting. Thus the magnitude of E is often used as a criterion to distinguish between diffusion-controlled and chemical reaction-controlled mechanisms. Low activation energies are indicative of diffusion-controlled kinetics, whereas high activation energies are indicative of chemical reaction-controlled kinetics. Ogwada and Sparks (1986a,b,c) investigated the effect of mixing on the energy of activation for ion exchange. Under static or low agitation where diffusion-limiting kinetics was dominant, activation energies were low but increased sharply under vigorous mixing where diffusion was no longer limiting.

SUMMARY

An overview of several models that are commonly used for the description of the retention of heavy metals in soils was presented. Single-reaction models were classified into equilibrium and kinetic types. A general-purpose multireaction kinetic and transport model was also presented. Major features of multireaction models are that they are flexible and are not restricted by the number of solute species present in the soil system nor by the governing retention reaction mechanisms. This includes reversible and irreversible reactions of the linear and nonlinear kinetic types. Moreover, these models can incorporate concurrent as well as consecutive-type retention reactions that may be equilibrium or kinetic in nature. Ion exchange mechanisms of the instantaneous and kinetic types were also presented. Rigorous validation of such models is needed for various contaminants and for soils having different physical and chemical properties. Model validation is a prerequisite step before model adoption as a predictive tool of the potential mobility of contaminants in soils.

REFERENCES

Abd-Elfattah, A. and K. Wada. Adsorption of Lead, Copper, Zinc, Cobalt, and Calcium by Soils that Differ in Cation-Exchange Materials, *J. Soil Sci.* 32, 271–283, 1981.

Amacher, M.C. Methods of Obtaining and Analyzing Kinetic Data, in *Rates of Soil Chemical Processes*, Sparks, D.L. and Suarez, D.L., Eds., *Soil Science Soc. Am. Inc.*, Spec. Publ. no. 27, Madison, WI, 1991, 19–59.

Amacher, M.C., H.M. Selim, and I.K. Iskandar. Kinetics of Chromium (VI) and Cadmium Retention in Soils: A Nonlinear Multireaction Model. *Soil Sci. Soc. Am. J.* 52, 398–408, 1988.

Amacher, M.C., H.M. Selim, and I.K. Iskandar. Kinetics of Mercuric Chloride Retention in *Soils. J. Environ. Qual.* 19, 382–388, 1990.

Amacher, M.C., J. Kotuby-Amacher, H.M. Selim, and I.K. Iskandar. Retention and Release of Metals by Soils: Evaluation of Several Models. *Geoderma.* 38, 131–154, 1986.

Aringhieri, R., P. Carrai, and G. Petruzzelli. Kinetics of Cu and Cd Adsorption by some Italian Soil. *Soil Sci.* 139, 197–204, 1985.

Barrow, N.J. Suitability of Sorption-Desorption Methods to Simulate Partitioning and Movement of Ions in Soils. *Ecol. Stud.* 74, 3–17, 1989.

Barrow, N.J. and T.C. Shaw. Effects of Solution and Vigor of Shaking on the Rate of Phosphate Adsorption by Soil. *J. Soil Sci.* 30, 67–76, 1979.

Bittel, J.E. and R.J. Miller. Lead, Cadmium and Calcium Selectivity Coefficients of a Montmorillonite, Illite and Kaolinite. *J. Environ. Qual.* 3, 250–253, 1974.

Bond, W.J. and I.R. Phillips. Approximate Solution for Cation Transport During Unsteady, Unsaturated Soil Water Flow. *Water Resour. Res.* 26, 2195–2205, 1990.

Buchter, B., B. Davidoff, M.C. Amacher, C. Hinz, I.K. Iskandar, and H.M. Selim. Correlation of Freundlich K_d and n Retention Parameters with Soils and Elements. *Soil Sci.* 148, 370–379, 1989.

Cho, C.M. Convective Transport of Ammonium with Nitrification in Soil. *Can. J. Soil Sci.* 51, 339–350, 1971.

Cernik, M., P. Federer, M. Borkovec, and H. Sticher. Modeling Heavy Metal Transport in a Contaminated Soil. *J. Environ. Qual.* 23, 1239–1248, 1994.

Harter, R.D. Kinetics of Metal Retention by Soils: Some Practical and Theoretical Considerations. *Agron. Abstr.,* p. 177, 1984.

Harmsen, K. *Behavior of Heavy Metals in Soils.* Centre for Agriculture Publishing and Documentation. Wageningen, The Netherlands, 1977.

Hinz, C. and H.M. Selim. Transport of Zn and Cd in Soils: Experimental Evidence and Modelling Approaches. *Soil Sci. Soc. Am. J.* 58, 1316–1327, 1994.

Jardine, P.M., J.C. Parker, and L.W. Zelazny. Kinetics and Mechanisms of Aluminum Adsorption on Kaolinite Using a Two-Site Nonequilibrium Transport Model. *Soil Sci. Soc. Am. J.* 49, 867–873, 1985.

Langmuir, I. The Adsorption of Gases on Plane Surfaces of Glass, Mica and Platinum. *J. Am. Chem. Soc.* 40, 1361–1402, 1918.

Lapidus, L. and N.L. Amundson. Mathematics for Adsorption in Beds. VI. The Effect of Longitudinal Diffusion on Ion Exchange and Chromatographic Column. *J. Phys. Chem.* 56, 984–988, 1952.

Montero, J.P., J.O. Munoz, R. Abeliuk, and M. Vauclin. A Solute Transport Model for the Acid Leaching of Copper in Soil Columns. *Soil Sci. Soc. Am. J.* 58, 678–686, 1994.

Ogwada, R.A. and D.L. Sparks. A Critical Evaluation on the Use of Kinetics for Determining Thermodynamics of Ion Exchange in Soils. *Soil Sci. Soc. Am. J.* 50, 300–305, 1986a.

Ogwada, R.A. and D.L. Sparks. Kinetics of Ion Exchange on Clay Minerals and Soil. I. Evaluation of Methods. *Soil Sci. Soc. Am. J.* 50, 1158–1162, 1986b.

Ogwada, R.A. and D.L. Sparks. Kinetics of Ion Exchange on Clay Minerals and Soil. II. Elucidation of Rate-Limiting Steps. *Soil Sci. Soc. Am. J.* 50, 1162–1166, 1986c.

Parker, J.C. and P.M. Jardine. Effect of Heterogeneous Adsorption Behavior on Ion Transport. *Water Resour. Res.* 22, 1334–1340, 1986.

Rasmuson, A. and I. Neretienks. Migration of Radionuclides in Fissured Rock. The Influence of Micropore Diffusion and Longitudinal Dispersion. *J. Geophys. Res.* 86, 3749–3758, 1981.

Rubin, J. and R.V. James. Dispersion-Affected Transport of Reacting Solution in Saturated Porous Media, Galerkin Method Applied to Equilibrium-Controlled Exchange in Unidirectional Steady Water Flow. *Water Resour. Res.* 9, 1332–1356, 1973.

Selim, H.M. Prediction of Contaminant Retention and Transport in Soils Using Kinetic Multireaction Models. *Environ. Health Perspect.* 83, 69–75, 1989.

Selim, H.M. Modeling the Transport and Retention of Inorganics in Soils. *Adv. Agron.,* 47, 331–384, 1992.

Selim, H.M., B. Buchter, C. Hinz, and L. Ma. Modeling the Transport and Retention of Cadmium in Soil: Multireaction and Multicomponent Approaches. *Soil Sci. Soc. Am. J.* 56, 1004–1015, 1992.

Selim, H.M., J.M. Davidson, and R.S. Mansell. Evaluation of a 2-Site Adsorption-Desorption Model for Describing Solute Transport in Soils. Proc. Summer Computer Simulation Conf. Washington, D.C. July 12–14, 1976, Simulation Councils Inc., La Jolla, CA, 1976, 444–448.

Selim, H.M. and I.K. Iskandar. 1981. A Model for Predicting Nitrogen Behavior in Slow and Rapid Infiltration Systems. In Modeling Wastewater Renovation — Land Treatment, I.K. Iskandar (ed.). John Wiley & Sons, NY, pp. 478–507.

Selim, H.M. and M.C. Amacher. A Second-Order Kinetic Approach for Modeling Solute Retention and Transport in Soils. *Water Resour. Res.* 24, 2061–2075, 1988.

Selim, H.M., M.C. Amacher, and I.K. Iskandar. Modeling the Transport of Chromium (VI) in Soil Columns. *Soil Sci. Soc. Am. J.* 53, 996–1004, 1989.

Sparks, D.L. *Kinetics of Soil Chemical Processes.* Academic Press, San Diego, CA, 1989.

Sposito, G. *The Surface Chemistry of Soils.* Oxford University Press, New York, 1984.

Starr, J.L. and J.-Y. Parlange. Dispersion in Soil Columns: The Snow Plow Effect. *Soil Sci. Soc. Am. J.* 45, 448–450, 1979.

Theis, T.L., R. Iyer, and L.W. Kaul. Kinetic Studies of Cadmium and Ferricyanide Adsorption on Goethite. *Environ. Sci. Technol.* 22, 1032–1017, 1988.

van der Zee, S.E.A.T.M. and W.H. van Riemsdijk. Sorption Kinetics and Transport of Phosphate in Sandy Soil. *Geoderma.* 3, 293–309, 1986.

CHAPTER **14**

Measurements and Modeling of Benzene Transport in a Discontinuous Permafrost Region

Larry D. Hinzman, Ronald A. Johnson, Douglas L. Kane, Ann M. Farris, and Greg J. Light

CONTENTS

PROJECT SUMMARY

Major objectives of this effort were to provide a quantitative description of the effects of discontinuous permafrost on subsurface hydrology and contaminant transport in the subsurface system at Fort Wainwright, Alaska (Figure 14.1), where significant petroleum releases had occurred. This description involved the use of SUTRA (Saturated-Unsaturated Transport), a numerical model that allowed us to predict the likelihood of various contaminant pathways. To accomplish these objectives, we initiated a series of data collection efforts including:

- Procurement of groundwater samples for organic and inorganic water chemistry parameters
- Measurement of water surface elevations
- Collection of meteorological and hydrologic data
- Characterization of the soils
- Injection of a tracer followed by subsequent sampling and analysis of the groundwater.

We shared these data with other participants, including the U.S. Geological Survey (USGS) and the U.S. Army Cold Regions Research and Engineering Laboratory (CRREL), and utilized data they collected including permafrost distribution information from CRREL and hydrologic data from the USGS.

We used all of the above information to help us in developing a site conceptual hydrogeological model and calibrating our contaminant transport model (SUTRA). The former is essential to the development of any groundwater flow or contaminant transport model. Phenomena such as recharge magnitude and preferential flow through permafrost-free zones can have dramatic effects on the fate of pollutants as predicted by a numerical model. In addition, the data used in calibration can

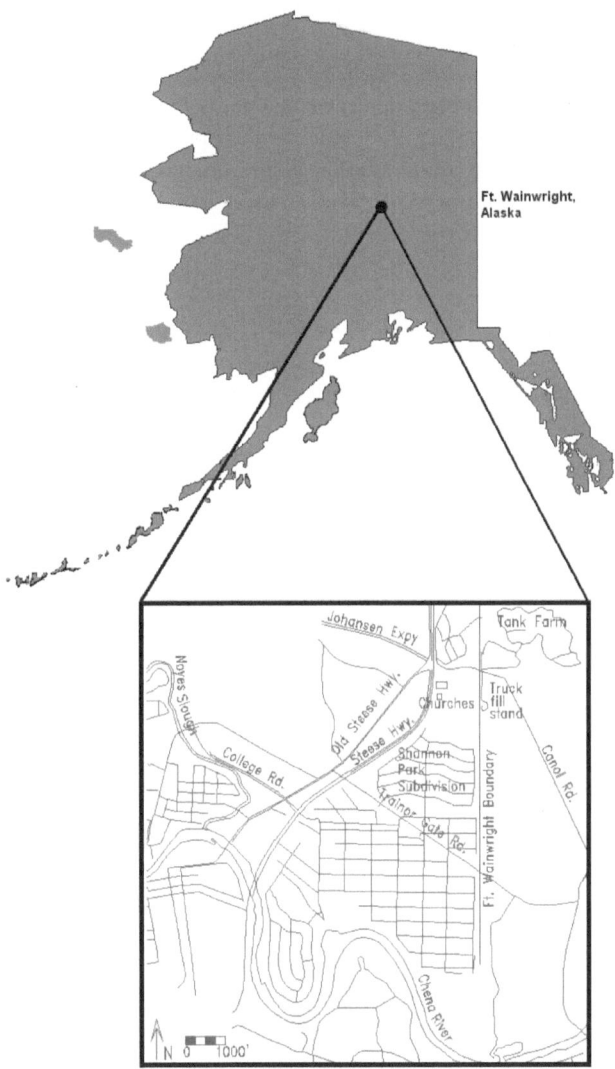

Figure 14.1 Vicinity map.

significantly affect the accuracy of the model's predictive capabilities. We, there-fore, invested extensive effort determining the accuracy of our chemical analyses and assessing the variation that can be introduced during sample collection.

To evaluate accurately the groundwater quality, the variability of field mea-surements must be considered. The effectiveness of an open bailer and an inertial pump in extracting representative groundwater samples from monitoring wells was compared in laboratory tests, using benzene as the contaminant. The bailer was most effective when the benzene concentration surrounding the well screen was uniform; bailer performance decreased as spatial variation in benzene concentration increased. However, the bailer did prove more effective than the inertial pump.

The laboratory tests were followed by field tests conducted at Ft. Wainwright. In samples collected with the bailer near the contaminant source, benzene concentrations tended to be significantly greater in the upper portion of the bailer than in the lower. Significant seasonal variation in benzene concentrations were also noted and attributed to fluctuating water levels.

Once the data analysis was completed, the concentrations from the tracer studies were used to estimate dispersion coefficients, hydraulic conductivities, and groundwater velocities in a permafrost-free channel. These values were used as input for the model simulations of the Tank Farm and Truck Fill Stand (TFS) source area on Ft. Wainwright (Figure 14.2). These simulations used benzene as the contaminant and were qualitatively calibrated using the field measurements of benzene concentrations described above.

The model results showed permafrost to act as a significant retarding force in contaminant transport to the point of stalling movement completely in some cases. Discontinuous masses of permafrost redirected the groundwater and created steeper gradients, thus changing the local hydrology significantly. Sensitivity analyses indicated concentrations were most affected by the source strength and the permafrost configuration. These results demonstrate the strong influence of permafrost on the hydrology and transport pathways.

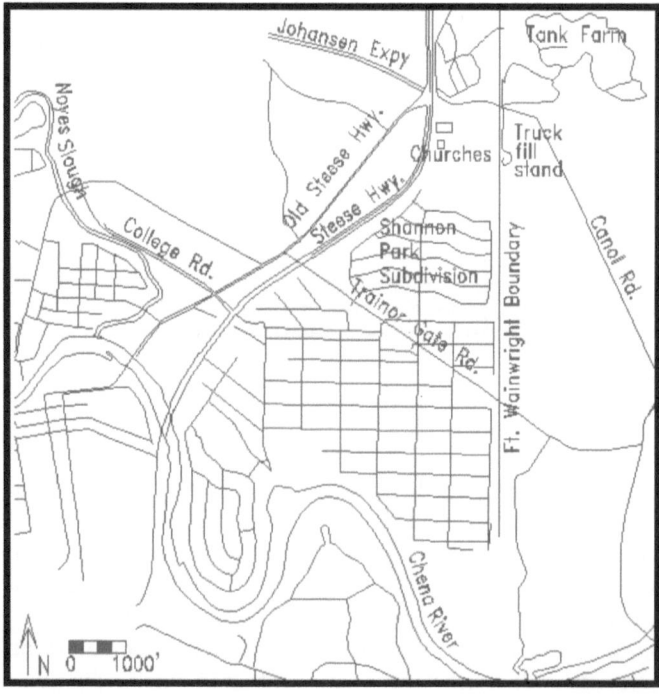

Figure 14.2 Site map.

PROBLEM DEFINITION AND SUPPORTING RESEARCH

There have been extensive spills of fuel oil, chlorinated solvents, and many other unspecified contaminants at nearly 100 locations on Fort Wainwright. Concern regarding contamination transport prompted a substantial preliminary data collection effort and work was initiated to develop an operational hydrologic model of subsurface flow rates and directions. Hydrology in discontinuous permafrost is extremely complicated and many obstacles were encountered in attempts to model transport plumes. Premier among these were the questions of subsurface flow processes. Understanding these processes is complicated by low hydraulic gradients that change direction (in some locations) and magnitude throughout the year and by the great spatial variability in subsurface media properties.

The Water and Environmental Research Center (WERC) at the University of Alaska at Fairbanks (UAF), in collaboration with the USGS and CRREL, conducted an analysis of the hydrology and contaminant transport at Fort Wainwright. Groundwater hydrology and contaminant transport has been an active science for several decades and has evolved into a sophisticated discipline. These processes are less well understood in permafrost regions and represent major obstacles to predicting contaminant pathlines in regions of discontinuous permafrost. UAF investigated the impact and effects of discontinuous permafrost on subsurface flow processes to enable successful prediction of contaminant transport at sites on Fort Wainwright. Our study encompassed two primary investigation: (1) an investigation of the effect of discontinuous permafrost upon subsurface hydrologic processes; and (2) the impact of local conditions (geology and permafrost distribution) on contaminant transport.

The specific purpose of this project was to determine the response of groundwater and contaminants to discontinuous masses of permafrost. The permafrost distribution in this area is quite complex and it would be virtually impossible to determine routes of chemistry pathlines from measurements alone. The model was an essential tool not just for predicting future contaminant levels, but also for mapping present contaminant pathlines. The contaminants follow directed and not particularly diffuse routes from their sources; consequently selection of well placement was both highly important and difficult.

Objectives and Scope of Work

One primary objective was to quantitatively describe the effect of discontinuous permafrost on subsurface hydrology. We compiled previously collected data on permafrost distribution in the specifically bounded area around the fuel terminal area. This included information determined from borings made by the Army Corps of Engineers (ACOE) and results obtained from the ground-penetrating radar studies conducted by CRREL personnel. These data were analyzed in concert with strategically located monitoring wells to indicate heads in permafrost-free areas and in sub- and suprapermafrost areas. Through the use of existing monitoring wells, we quantitatively described the annual dynamics of subsurface flows incorporating the influence of these complicated subsurface properties.

A second major goal was to characterize the effect of the local conditions including discontinuous permafrost on the fate and transport of hydrocarbon contaminants in groundwater. Our approach was to review the existing site-specific data to ascertain what was known with respect to the sources of contamination, major constituents contributing to contamination, groundwater quality, immiscible phase distribution in both the unsaturated and saturated zones, and estimated spill history. We performed a literature search to delineate critical properties of suspected major contaminant species, including surface tension, diffusivity, viscosity, solubility, and partition coefficients. We identified several significant gaps in the data, which needed to be filled to properly assess the extent of contamination and provide a prognosis for the future extent. These gaps included critical constants needed for quantifying adsorption isotherms and partition coefficients. Both laboratory (soil carbon) and field (tracer) studies were conducted to fill some of these data gaps. The objective of the tracer study was to provide solid information upon parameters such as dispersivity and velocity. Retardation depends on both the site-specific soil organic carbon content and the organic/water partition factor; we measured the former at selected sites and calculated the latter. We focused particular attention upon concentrations of benzene in the groundwater because the transport of this solute generally presents the major problem with respect to adverse effects on the environment, including drinking water quality.

Our approach was a rigorous field investigation based upon measurements of water table elevations, water chemistry, soil thermal conditions, and local meteorologic conditions. A substantial set of piezometers previously existed in the study area. These were supplemented with additional nested piezometers critically placed at permafrost boundaries to establish vertical flow gradients. These water levels were measured in cooperation with the USGS. Water samples were collected monthly during the summer and once every 3 months during the winter. Characterization of surface conditions was accomplished through physical measurements of soil temperature and moisture content, precipitation, air temperature, and relative humidity. During spring thaw, snow surveys were conducted daily to measure rates of snowmelt, thus enabling us to characterize groundwater recharge rates and locations.

Numerical models were utilized to simulate these important subsurface processes. SUTRA was used to simulate solute transport and the hydrologic processes. Solute transport is a coupled process consisting of the natural groundwater flow and the transport of dissolved matter. The accuracy of the model predictions of solute transport depends upon accurately determining the model parameters related to both processes. Tracer studies were needed to develop reliable estimates of these essential model parameters. The primary value of a tracer study is an accurate measurement of flow velocities and directions with very small error and an actual measurement of the flow pathlines. In no other effort in the permafrost environment, either previously or concurrently, did anyone attempt to directly measure the actual pathlines of subsurface contaminants. This was one of the most critical pieces of information of the entire program. Another benefit was measurements of dispersivity. The tracer studies did more than provide values for these parameters; the studies provided an explanation of the contaminant behavior under the site-specific conditions.

Dispersivity is a scale-dependent indicator of the degree of mixing which does not lend itself to direct measurement. This process normally consists of estimating the parameters appearing in the governing equation(s) from observations made on the dependent variables. We began this process by employing analytical one- and two-dimensional solutions that were solved either analytically or numerically for longitudinal and transverse dispersion coefficients. For example, in a one-dimensional approximation for slug injection of a known mass, the concentration depends upon a dimensionless time and the Péclet number. From concentration profiles obtained over time at the monitoring wells, we could solve the governing equation for both the pore velocity and longitudinal dispersion coefficient that produced the best match with the data.

Project Location and Background

Fort Wainwright is located at the eastern end of Fairbanks, Alaska. The Tank Farm and TFS source areas, located in Operable Unit 3 (OU3) on the post, were the focus of this investigation. These areas contain large patches of near-surface permafrost and significant petroleum contamination and have been the focus of numerous other studies that provided background data. The areas are located north of the Chena River, along the northwestern border of the post, at the base of Birch Hill. Specific components include:

- 14 empty 420,000-gal aboveground storage tanks (ASTs)
- 2 empty 1,000,000-gal ASTs
- Building 1173 — garage/office — 2 underground storage tanks (USTs)
- Building 1183 and 1182 — terminal/manifold
- Warehouse
- An 8-in. Fairbanks–Eielson pipeline runs east of Tank Farm to Building 1183
- Three 8-in. and four 3-in. CANOL pipelines run along CANOL Service Road through Valve Pit A
- Two 3-in. pipelines leading from the Birch Hill ASTs to the Truck Fill Stand (TFS) area
- TFS area — 2 ASTs

The 14 420,000-gal ASTs located on Birch Hill were installed in 1943 and were expected to be in use for only 5 years, but were in use for over 45 years, until they were decommissioned in 1990 (Ecology and Environment, Inc., 1994). Approximately one third of the known spills have occurred from these tanks and the two 1,000,000-gal ASTs that are also located on the hill. All 16 tanks were decommissioned by September 1993 (Ecology and Environment Inc., 1994). The other major area of known spills has been the TFS area. The ASTs in the TFS, the USTs by the buildings, and the ASTs on Birch Hill were considered in this study to be included in and defined as the Tank Farm source area. Quantities of fuel spilled, exact dates of occurrence, and even locations are unknown for many of the spills. The sum of the known spill quantities is approximately 90,000 gal. This represents 20% of the 450,000 gal estimated by CRREL from field measurements (Currier et al., 1994). It is clear that many spills have gone completely unrecorded or undetected. A probable source of contamination that has remained almost completely undocumented was the standard practice of draining water from the bottom of the tanks into the surrounding dike to either evaporate or infiltrate into the soil (McCort, 1994).

The specific geology and hydrogeology of the Tank Farm area was described in Ecology and Environment, Inc. (1994). They separated the site into three areas: (1) Birch Hill; (2) the base of Birch Hill; and (3) the floodplain, located between the base of the hill and the Chena River. Birch Hill consists mainly of Birch Creek Schist with 0 to 30 ft of windblown deposits of Fairbanks Loess on top (Ecology and Environment, Inc., 1994). No permafrost has been noted on this south-facing hill. The base of Birch Hill is a transition zone between the schist and the Chena Alluvium. There are variable thicknesses of these materials in this area as well as Fairbanks Loess (Ecology and Environment, Inc., 1994). Near-surface permafrost has been noted in this area in both soil borings and the Ground Penetrating Radar (GPR) work (Lawson et al., 1994). The floodplain consists mainly of Chena Alluvium overlying deep — at least 100 ft below ground surface (bgs) — Birch Creek Schist (Ecology and Environment, Inc., 1994). Extensive permafrost has also been noted (see Figure 14.3A) throughout the floodplain. Indications of an old slough are noted, extending west from CANOL Road, approximately 525 ft south of the Tank Farm fence, through the TFS area toward Lazelle Road (Lawson et al., 1994).

UAF's work on this project was conducted in cooperation with CRREL and the USGS. The USGS had an ongoing project to create and maintain a subregional groundwater simulation of Fairbanks using the USGS Modular Finite Difference Groundwater Flow (MODFLOW) computer code. The simulation grid includes boundaries on the north and west defined by the bedrock hills of the Yukon–Tanana Uplands. The southern boundary is the Tanana River, and constant head values interpolated from the Chena and Tanana River stages define the eastern boundary. Simulated hydraulic conductivities range from 250 ft/day in the flood plain, to 5 ft/day in the silt, to 0.1 ft/day in the bedrock, to 0.001 ft/day in permafrost. The area surrounding the Tank Farm has a recharge input of approximately 2 in/year in the hills and 7 in/year on the floodplain (Claar, 1995). Model analyses using the groundwater model MODFLOW indicated that the vertical conductivity is approximately 1/15 the horizontal conductivity for the same soil type (Claar, 1995).

The USGS also simulated the bedrock areas in and around Birch Hill using MODFLOW (Bolton, 1996). These simulations were designed to explore the hydrologic connection between the bedrock hills and the alluvial floodplain. The results showed that the total flow from the bedrock aquifer was several orders of magnitude less than the total flow in the floodplain. The USGS results also indicated minimal upward vertical flow from the bedrock beneath the alluvium into the silts and gravels that overlay it (Bolton, 1996).

A Chena River head profile, used as a boundary in both the USGS subregional model and in the SUTRA simulations, was constructed using the mean annual stage values for April 1994 to 1995 (Claar and Bolton, 1996). Mean stages were based on data collected from continuous recorders and monthly measurements at 12 sites along the Chena River, and the potentiometric surface in river cells between the sites was obtained through linear interpolation.

Lawson et al. (1994) did work with ground penetrating radar (GPR) and borehole data to construct a map of the permafrost in the OU3 area. These maps were used in all the SUTRA simulations as a basis for determining possible flow pathways and hydraulic conductivities in the aquifer (Figures 14.3A and 14.3B).

Currier et al. (1994) completed a Microwell® investigation in order to delineate the extent of floating product on the water table in the Tank Farm area. Starting at the base of Birch Hill in the Tank Farm, small diameter, drive-point wells (Microwells®) were driven and simultaneously sampled for chemical analysis. Wells were extended out in each direction until no contamination was found. They discovered an estimated 4400 gal of free product and an area of 100,000 ft² of product sheen (Figure 14.4). Assuming a soil porosity of 0.3 and residual saturation of 15%, they estimated 450,000 gal of product was being held in the soil.

Site Conceptual Hydrogeological Model

The correct conceptualization of the hydrogeologic system is critical in development of any groundwater model. How one defines the surrounding boundary conditions or the internal structure of the geology, geomorphology, or media properties significantly influences what results the model will produce. In the OU3 area, the presence of ancient sloughs, discontinuous permafrost, low hydraulic gradients, and the rise and fall of the Chena River make this system very complex. Our conceptualization of this system is based upon information of the permafrost distri-

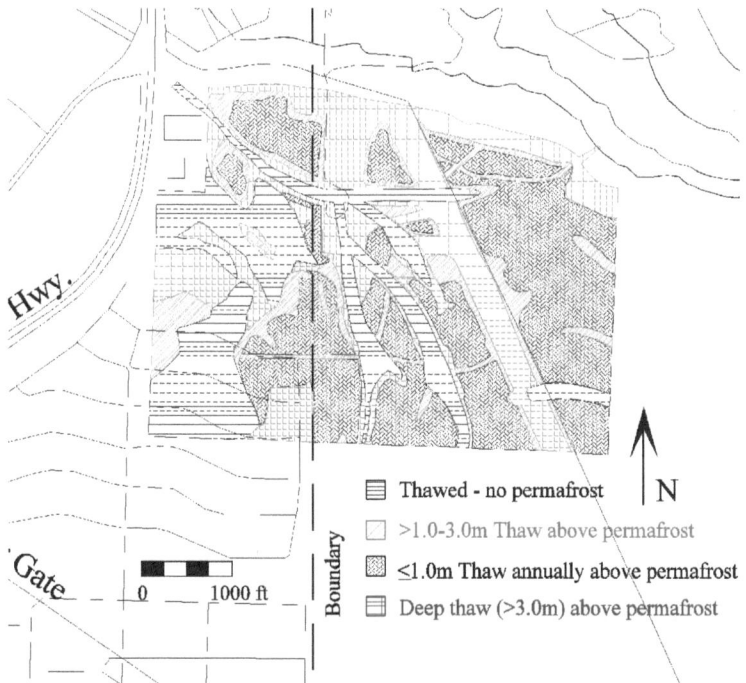

Figure 14.3A Permafrost configuration (From Lawson et al., 1994).

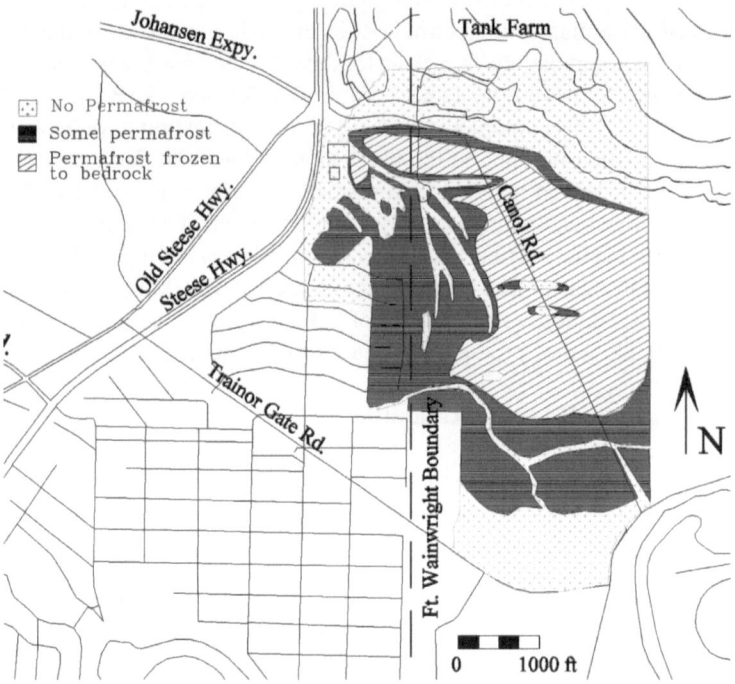

Figure 14.3B Permafrost configuration (From Lawson et al., 1994).

bution compiled by CRREL (Lawson et al., 1994), work by Ecology and Environment, Inc. (1994), and our own extensive observations on the site. Even though this analysis was directed at a relatively small area (a portion of OU3), one may make the analogy of the aquifer to a piece of paper, in that the vertical dimension (depth) is very thin (<100 ft) when compared to the horizontal dimension (5000 ft). An important aspect of this analysis is the necessity of separating the subpermafrost flow system from the suprapermafrost flow system.

Combining groundwater elevations collected throughout the study and permafrost distribution data collected by CRREL, it appears that the water levels in permafrost-free areas never rose above the level of the top of the adjacent permafrost. This has important consequences in understanding the behavior of suprapermafrost groundwater. Although the permafrost will act as a confining layer and will hold a perched water table, water on top of the permafrost will flow to the permafrost-free areas where it may enter the subpermafrost system. This not only controls where recharge will occur but also impacts solute movement. In light of the above, it is important to correctly discern the difference between the water table in the permafrost-free areas vs. the perched water table and the potentiometric surface. The potentiometric surface can be displayed by water elevation contours, which extend across the area regardless of where permafrost exists. The surface represents the gradients of water potential indicating the directions in which hydraulic head increases or decreases, but not necessarily the direction in which water will flow. If there were no vertical flow components at

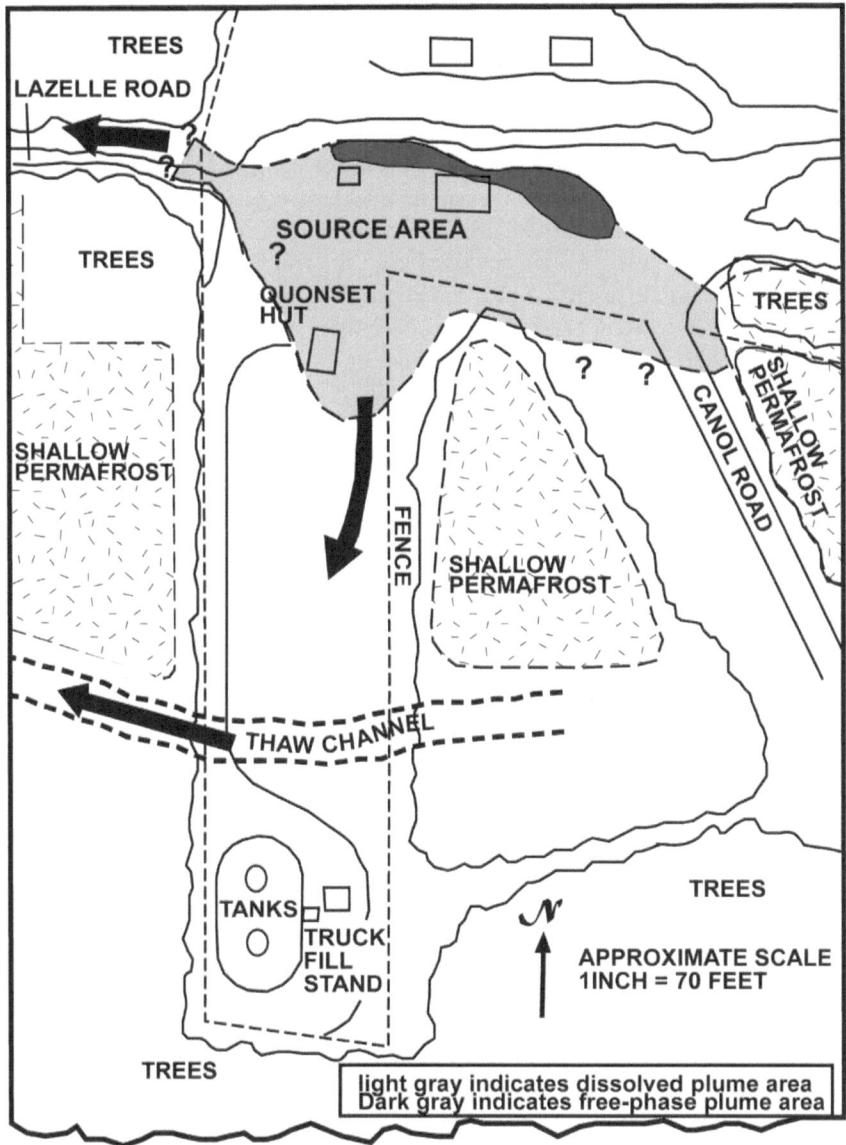

Figure 14.4 CRREL source delineation (Currier et al., 1994).

the measuring point where the well was installed, the potentiometric surface would be the same as the water table. Water flows from areas of higher potential to areas of lower potential, but it is directed by the bounding permafrost. It appears that the permafrost extends completely to the bedrock in the central portion of our study area. This is analogous to a long wall that would act to channel flow longitudinal to the wall, restricting flow and the spread of contaminants to the south of this wall. The essential conclusion in this interpretation is that **flow above permafrost does not**

represent a major analysis factor in determining contaminant pathlines unless the source is over permafrost. Suprapermafrost flow will tend to the permafrost-free areas where it will either move with the unconfined groundwater in the permafrost-free areas or it will interact with the subpermafrost system. This behavior allows us to proceed with two-dimensional numeric analyses. The third dimension is included by changing the thickness of the aquifer. Where the permafrost extends from above the static water table to bedrock, the aquifer thickness goes to zero.

Accurate prediction of contaminant movement requires much more than selecting a model and running it. As discussed previously, the conceptual model will to a great extent define what the results will be; consequently, it is essential to critically evaluate the conceptual model and validate it while calibrating and verifying the parameterization. As one step in evaluating our conceptual model, we needed to confirm that the groundwater regime north of the permafrost "wall" is indeed distinct from the south side.

Subsurface media properties vary substantially across the study area. Soils were developed through erosional and depositional processes of the Chena and Tanana Rivers and through eolian deposits of loess. Ancient oxbow sloughs contain very fine-grained deposits of silt or perhaps clay adjacent to very coarse-grained sand and gravel alluvium deposits. This heterogeneity in soil conditions presents difficulties in interpretations or estimations of values of porosity, hydraulic conductivity, and dispersivity. Estimation of these parameters was based upon interpretation of all available well logs, geologic investigations, and results of tracer studies.

Project Approach

Our goal was to develop an understanding of the flow regime in discontinuous permafrost, identify the important processes controlling rates of flow, and incorporate them into algorithms that may be used as a guide for modeling contaminant transport anywhere in discontinuous permafrost. In order to develop a full understanding, we must proceed in a rational progression from observation to simulation. We needed to first understand the macroscale processes before we could understand the microscale processes.

The steps to modeling the contaminant movement on any site are these:

1. Study the present hydrogeologic system to develop an understanding of what controls the system and where those controls are acting.
2. Define a conceptual model that incorporates all of those controlling factors to allow an accurate simulation.
3. Define the parameterization in a computer program to emulate the conceptual model which describes the actual system.
4. Execute the model.
5. Compare the simulated results with measured conditions.
6. If the model does not compare well, determine if the conceptual model is inadequate for some reason or if parameter adjustment is needed.
7. Calibrate model. This step ensures that the model can reproduce field-measured heads, flow rates, and solute concentrations. There is an uncertainty in defining the exact spatial and temporal distribution of parameter values. We conducted a sensitivity analysis to determine the effect of this uncertainty.

Hydrologic Data Collection

Water levels were measured at Fort Wainwright on a monthly basis, except in winter. All wells were measured within a 3-day period during the winter and a 1- or 2-day period in the summer. The technique for normal water table wells in permafrost-free areas is typical of similar measurements in more temperate regions. A piezometer was opened/unlocked and an electronic tape was used to determine the exact depth to the surface of the water. Measurements were repeated until consecutive readings were within 0.01 ft. Wells through permafrost must be pressurized with air to force the water out of the bottom of the piezometer to prevent it from freezing and rendering the well unusable. After the well was depressurized, the water was allowed to equilibrate. When consecutive water level measurements within a few minutes were within 0.01 ft of each other, we considered the system at equilibrium. After measuring the water level, the wells were repressurized with oilless breathing air. The pressure head applied to each well depended upon the amount of water above the bottom of the well:

$$P = h \, \gamma \qquad (14.1)$$

where   P = pressure
        h = height of water column above screen
        γ = specific weight of water

After the water level was measured, the electronic tape was decontaminated.

Meteorologic Data Collection

The surface represents an additional boundary of the regional and site-specific models for which fluxes must be specified. The hydrologic processes that must be quantified are rainfall, evaporation, snowmelt, runoff, and infiltration. The thermal processes are those which affect ground freezing and thawing. Evaporation and snowmelt are hydrologic processes that are most easily quantified through energy analyses. To insure our capability to quantify these processes, meteorologic data were collected. Figure 14.5 shows the instrumentation installed at Fort Wainwright.

The necessity of obtaining continuous meteorologic data for regional modeling has been previously described. Knowledge of surface flux is equally critical for the site-specific solute transport modeling. Correctly identifying subsurface flow path-lines of contaminants is completely dependent upon correctly quantifying the hydrologic portion of the model. To this end, we must be able to describe the time series processes of rainfall and snowmelt infiltration. In order to accomplish these tasks, we must be able to describe the abstractions due to evaporation and soil moisture storage and the limitations to infiltration due to frozen ground.

The thermal condition of the near-surface soils (top 10 m) has a major impact upon contaminant transport processes in several aspects. The most important aspect is whether the surface is frozen or thawed and to what depth. This can be modeled quite easily given information on the surface energy balance. The subsurface

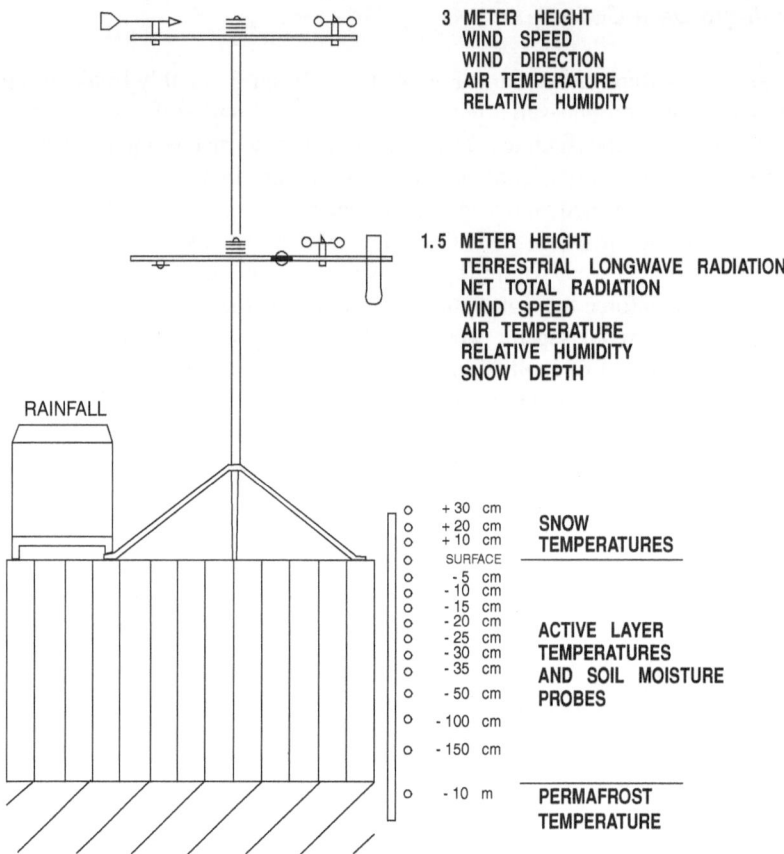

Figure 14.5 Meteorological instrumentation at Fort Wainwright.

temperatures also impact the rate of water movement through changes in viscosity and will impact the subsurface chemistry reactions and retardation. The subsurface temperatures can also provide additional information on the source of the groundwater and, when coupled with adequate information on basic groundwater geochemistry and flow regime, allows analyses of where the recharge occurred.

Two meteorological stations were located within the boundaries of the Tank Farm with instrumentation sufficient to permit detailed analysis of the surface energy balance for an undisturbed site. Meteorological parameters measured include air temperature, relative humidity, precipitation, wind velocity, wind run, wind direction, soil and snow temperatures, and the radiation components. Radiation measurements include incoming shortwave, emitted longwave, and net radiation absorbed. Soil temperatures were measured and recorded hourly. Unfrozen soil moisture was measured using a Tektronix Time Domain Reflectometer (TDR) and recorded every 3 h. These measurements were recorded on Campbell Scientific CR10 dataloggers. Radiation was measured with a horizontal sensor 1.5 m above the soil surface. All radiation sensors were calibrated annually in our laboratory and are sent to the Eppley Laboratory for recalibration as needed.

Incident solar radiation was measured using an Eppley double-dome spectral pyranometer. The spectral range of this sensor was 0.3 to 3 μm, which excludes longwave terrestrial radiation. The accuracy of each sensor is reported as ±1%. The cosine response was less than 1% when the sun angle was within 0 to 70° of perpendicular of the sensor plane. Longwave radiation is measured with an Eppley precision infrared pyrgeometer. The spectral range of this sensor is between 4 and 50 μm and the cosine response was better than 5% from normalization. The net absorbed radiation (0.3 to 60 μm) was measured using a Met One Q6 net radiometer. This sensor measures the total radiation absorbed.

We used a Campbell Scientific model 207 or model HMP 35C temperature and relative humidity probes at each operating site to determine the temperature and humidity gradients. Campbell Scientific reports worst-case accuracy of ±0.4°C between −33 and +48°C for the temperature sensor, and 3% error between 12 and 100% relative humidity. These probes were housed in a self-aspirating radiation shield. Mean wind speed, mean wind vector direction, and standard deviation of direction were recorded hourly. Wind velocity, wind run, and wind direction were measured using an R M Young Model 05305 Wind Monitor-AQ at the lower met site. The threshold of wind measurement was 0.4 m/s. The accuracy is ±25% linearity. Station pressure was measured using a Vaisala PTA 427 pressure transmitter. Reported accuracy is ±0.4 mb. Wind run and precipitation were totaled continuously. All other meteorological parameters were measured every minute. In addition to the above measurements, the snowpack water equivalent was measured periodically during snowmelt with an Adirondack snow sampler. Due to the variability of the snowpack between the hillsides and the floodplain, the accuracy of this technique is difficult to quantify (Rovansek et al., 1993).

The snow water content was measured at many locations prior to initiation of spring snowmelt and daily during melt. In 1994 and 1995, the snow was quite uniformly distributed across the area surrounding the TFS and across the southern slope of Birch Hill. The initial snowpack water content averages for 1994 and 1995 were 9.3 and 13.1 cm (Plumb and Lilly, 1996), respectively. In April 1994, the Fairbanks Field Office of the Natural Resources Conservation Service (NRCS) reported a snow water equivalent of 9.9 cm. Their April 1995 value was 13.2 cm. These both compare to a 30-year average of 9.9 cm (NRCS, 1994, 1995). These data indicate that the snowpack of 1994 was quite close to the long-term average snowpack, while 1995 was significantly higher than average. This also indicates that snow measurements collected by NRSC may be applied to the Fort Wainwright area, since the snowpack around Fairbanks does not experience major redistribution by the wind as experienced in the more windy parts of Alaska. A snow survey of snowpack depth and water content was collected at 10 points daily during snowmelt.

Air temperatures were measured at the primary site at the base of Birch Hill and at a secondary site near the top of Birch Hill. The annual average temperatures at the primary site were compared with annual averages and long-term mean temperatures collected at Fairbanks International Airport (FIA). The 1994 and 1995 average air temperatures were −3.7 and −2.5°C respectively while at the FIA, the annual air temperatures were −2.6 and −2.1°C. The 30-year average temperature at FIA was

–2.8°C. Liquid precipitation was measured at the primary meteorologic station and at the Birch Hill site. Examination of the annual total precipitation measured at FIA shows that both 1994 (246 mm) and 1995 (222 mm) were somewhat drier than the 30-year average of 276 mm. Summer rainfall (June through September) at the Fort Wainwright station was 128 and 152 mm in 1994 and 1995, respectively, while at FIA, we received 139 and 170 mm in 1994 and 1995. The 30-year average for the summer period is 156 mm, demonstrating that, in terms of summer rainfall, both years were reasonably close to the long-term average. Relative humidity measured at the two sites shows typically good comparison. Differences arise due to low cloud ceilings and, in winter, during atmospheric inversions.

Three components of the radiation balance (incident shortwave, terrestrial long-wave, and net radiation) were measured at the primary meteorologic site. These three measurements are most important in terms of enabling calculations of snow-melt and evaporation, both of which are significant factors in hydrologic analyses. An assumption that the surface albedo is equal to 0.8 while the ground is covered with snow, and 0.2 during snow-free conditions enables calculation of the other two components of the radiation balance (reflected shortwave and atmospheric longwave radiation).

Evapotranspiration (ET) may be calculated from these simple data using one of several formulations (Kane et al., 1990). In this study, ET was estimated follow-ing the method of Priestley and Taylor (1972), which relates potential evaporation to net surface energy flux. Net surface energy flux is calculated using a simplified surface energy balance and relates actual ET to potential evaporation by an empir-ical parameter α.

$$Q_e = \alpha \, s/(s + h) \, (Q_{net} - Q_c) \tag{14.2}$$

where

Q_e = energy available for evaporation (W/m^2)
α = ratio of actual to potential ET
s = slope of the specific humidity and temperature curve (°C^{-1})
h = psychrometric constant in terms of specific humidity (°C^{-1})
Q_{net} = net radiation (W/m^2)
Q_c = energy conducted into the ground (W/m^2)

The energy conducted into the soil was calculated based upon thermistors at 7 and 22 cm depth. Rouse and Stewart (1972) and Stewart and Rouse (1976) found, in a subarctic setting, that they could describe $(s/s + h)$ as a linear function of screen air temperature.

$$s/(s + h) = 0.406 + 0.011(T_a) \tag{14.3}$$

where T_a is the air temperature (°C).

Calculated ET is included in Figure 14.6 plotted along with data on daily precipitation. ET is the dominant term of the surface water balance during the summer in interior Alaska (Gieck and Kane, 1986).

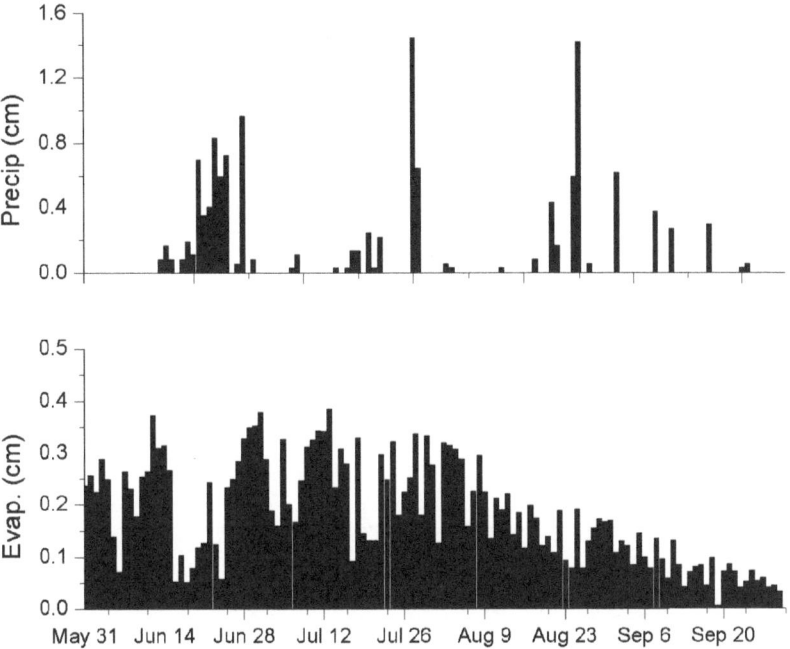

Figure 14.6 Calculated daily evapotranspiration during the 1994 summer.

Soils Measurements

Soil thermistors and TDR probes were installed in three sites around the Tank Farm. Soil temperatures were measured hourly in a soil profile to enable energy balance calculations. An automated TDR system, operated near the primary meteorologic station, collected measurements every 6 h during the summer months. Two sets of TDR probes were installed at this site: one on the hillslope of Birch Hill and the other on the valley floor. Probes, which could be periodically measured using a portable TDR, were installed at two secondary sites, one near the top of Birch Hill and the other just south of the TFS. These probes were measured infrequently but give an indication of near-surface moisture dynamics. The probes, which were measured automatically, demonstrate the integrated response of the surface soil to thawing or freezing and snowmelt or rainfall inputs and losses to ET or percolation to groundwater. The data from these probes can give reliable measurements of the changes in near-surface soil moisture. The total soil moisture of the profile (in cm of water) represented by the extent of the probes may be calculated as the sum of the moisture content (% by volume) times the incremental depth of soil measured by each probe. The change in soil moisture content (ΔSM) is then the difference from day to day. These values may be used to determine infiltration (I), given measurements of precipitation (P) and ET as calculated above.

$$\Delta SM = P - ET - I \qquad (14.4)$$

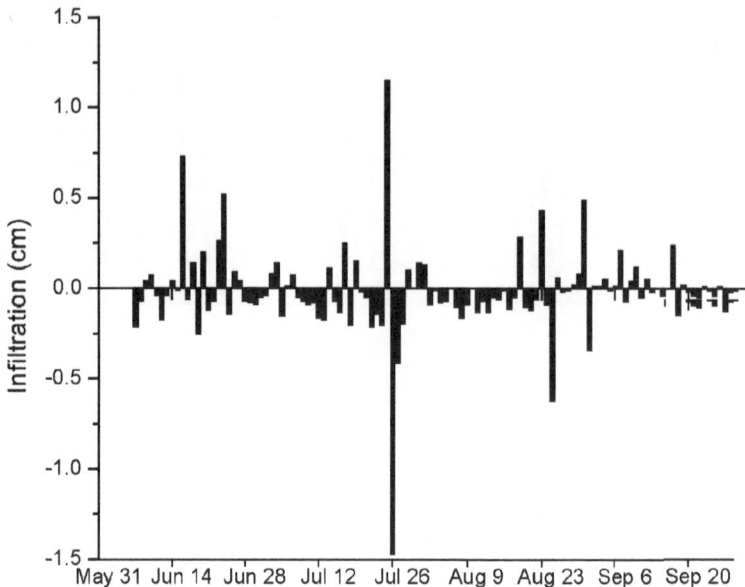

Figure 14.7 Daily rates of infiltration during summer 1994.

Daily rates of infiltration are shown in Figure 14.7. In interior Alaska, evaporative demands nearly equal precipitation during the summer months (Gieck and Kane, 1986), although more intense autumn storms often do provide some groundwater recharge. Most groundwater recharge occurs during spring snowmelt (Kane and Stein, 1983a).

FORT WAINWRIGHT BTEX SAMPLING AND BENZENE VARIABILITY STUDY

Between April 1994 and May 1995, we investigated groundwater transport of hydrocarbon contaminants in the Tank Farm and TFS areas on Fort Wainwright. The Tank Farm is the site of numerous petroleum releases since the 1960s and was the focus of the groundwater sampling aspect of this investigation.

Prior to our work, the Corps of Engineers (USACE) installed five groundwater-monitoring wells in 1987. These wells were sampled periodically through 1993 (U. S. Army Corps of Engineers, 1993). Ecology and Environment, Inc. (1993) was contracted to conduct a remedial investigation to delineate the extent of contamination on Fort Wainwright. Several more monitoring wells were installed in the vicinity of the Tank Farm as part of the Ecology and Environment, Inc. investigation, which confirmed the presence of benzene, toluene, ethylbenzene, and total xylene (BTEX) contamination in the soil and groundwater. Total BTEX contamination in the groundwater in OU3 was found to be as high as 24.1 ppm (Ecology and Environment, Inc., 1994).

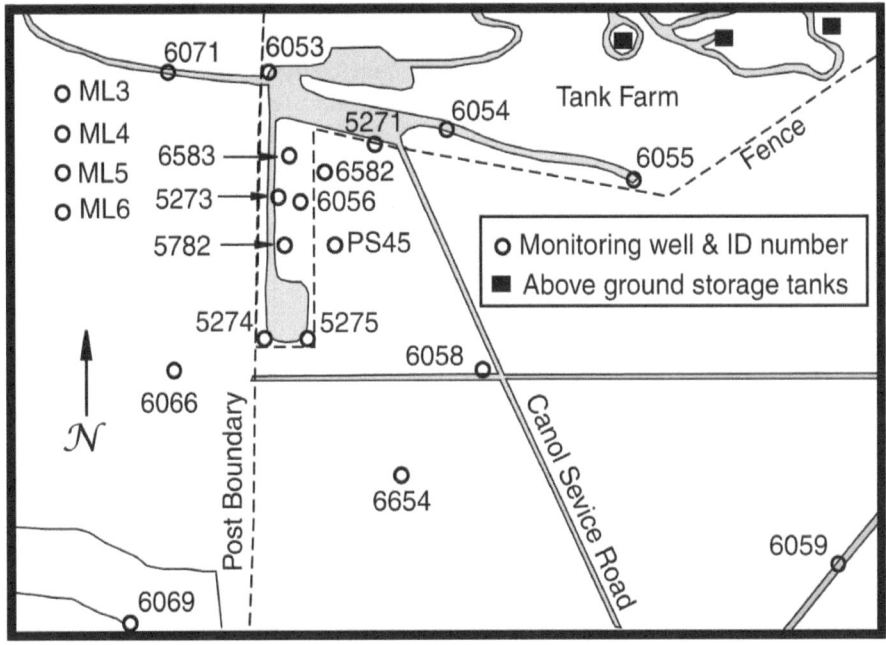

Figure 14.8 Highest benzene concentrations measured during our investigation at OU3 monitoring wells.

As part of our program, groundwater samples were collected for benzene concentration from existing and newly installed monitoring wells at the base of Birch Hill and in the Chena River floodplain (Figure 14.8). The monitoring wells used were screened between 6 and 10 m below grade. During the period in which samples were collected, water levels ranged between approximately 5.0 and 5.6 m below grade.

Sampling and Analysis

Prior to sampling, each monitoring well was purged using an Isco™ variable speed peristaltic pump. A 0.95-cm ID polyethylene tube was attached to the pump and the inlet was placed at the water/air interface inside of the well. The outlet tube was attached to a Hydrolab™, an instrument that constantly monitors pH, conductivity, temperature, and dissolved oxygen as the purge water passes over a probe. These parameters were recorded every 2 min. The well was purged until these parameters stabilized to within 10% of the previous reading, typically 10 to 15 min. During the purging process, the inlet tube was slowly raised and lowered to ensure that all of the static water in the well was removed. After purging was completed, sufficient time was given for the fine particles to settle out before sampling so the samples would not be turbid. The purge water was collected in properly labeled 55-gal drums and delivered to a designated site on post for treatment.

A Teflon bailer or an inertial pump was used to extract the samples from the monitoring wells. The bailer collection process was as follows: (1) the bailer was suspended by a nylon fishing line and slowly lowered, to avoid splashing the water and potentially causing volatilization of the contaminants, into the well; (2) the sample was drained from the bailer into a commercially cleaned 40-mL sample vial via a bottom-emptying device consisting of a narrow Teflon tube; and (3) the tube was placed into the bottom of the vial and water was allowed to fill up around the tube and overflow the vial for several seconds to minimize the volume of sample water contacting the atmosphere.

The inertial pump consists of a small check valve with a threaded end that is screwed into the bottom of a section of polyethylene tubing that was placed into the well so that approximately 2 m of tubing extended out of the well. The samples were obtained by repeatedly raising and lowering the tube, which forced the sample up and out of the tube. The sample was then collected in a similar fashion as with the bailer. The inertial pump was used only to sample the 2.54-cm ID drive point wells that were too narrow to be sampled with the bailer.

All groundwater samples were collected in commercially cleaned 40-mL vials equipped with Teflon-lined caps. The vials were filled so that a positive meniscus was formed and no air bubbles were trapped in the vials. The samples were acidified to a pH of less than 2 by the addition of several drops of 1:1 solution of concentrated hydrochloric acid and distilled water. The samples were immediately placed in an ice-filled cooler. The samples were kept in the cooler until delivered to the WERC laboratory where the samples were stored at 4°C. All samples were analyzed within 7 days.

Analysis was performed according to a slightly modified EPA Method 602 (*Federal Register*, 1984). The primary modification to the method is that a FID detector was used instead of a PID. The reason for this modification is that the GC made available for this study is equipped with a FID. The FID is widely used because of its high sensitivity to organic carbon-containing compounds (APHA et al., 1992). BTEX constituents were extracted from the samples using a Tekmar LCS-2 Liquid Sample Concentrator and sorbed onto a 25-cm Tekmar Tenex trap packed with 23 cm of 2,6-diphenylene oxide polymer. The carrier gas was nitrogen. The samples were purged for 12 min, desorbed from the trap for 4 min at 180°C, and the trap was then baked for 12 min at 275°C. The samples were then auto-injected into a Hewlett-Packard 5890 GC equipped with a FID. Nitrogen was the carrier gas at a flow rate of 12.4 mL/min. The injector temperature was 295°C, detector temperature was 300°C, initial oven temperature was 50°C, and the final oven temperature was 290°C.

Known additions of BTEX were added to field samples and analyzed to ensure that there were no substances in the groundwater matrix that would interfere with BTEX recovery when analyzed on the GC. Matrix spikes were used as a measure of analytical accuracy in the measurement of BTEX, as described in the Quality Assurance Program Plan (Currier and Leggett, 1994).

Results of BTEX Sampling

Of the 27 wells sampled, 6 (AP5271, AP5274, AP5273, AP6583, AP6582, and ML-3) contained benzene above the maximum contaminant level (MCL) for drinking water of 5 ppb (Tables 14.1 and 14.2). One sample collected from AP5271 (April

Table 14.1 Summary of BTEX Concentrations in Sampled Wells on Various Dates in 1994

AP Well #	Sampling Date	Benzene (ppb)	Toluene (ppb)	Ethyl Benzene (ppb)	Total Xylene (ppb)	Total BTEX (ppb)
5274	4/8/94	10.9	BDL	BDL	BDL	10.9
6053	4/8/94	BDL[a]	BDL[a]	BDL[a]	BDL[a]	BDL[a]
6054	4/8/94	4.0	BDL	BDL	BDL	4.0
6055	4/8/94	BDL	BDL	BDL	BDL	BDL
6058	4/8/94	BDL	BDL	BDL	BDL	BDL
6059	4/8/94	BDL	BDL	BDL	BDL	BDL
6069	4/8/94	BDL	BDL	BDL	BDL	BDL
5271	6/15/94	10.4	4.7	4.4	14.3	33.8
5273	6/15/94	67.1	1.2	BDL	1.7	70
5274	6/15/94	BDL	BDL	BDL	BDL	BDL
6053	6/15/94	BDL[a]	BDL[a]	13.3	3.9	17.2
6054	6/15/94	BDL	BDL	BDL	BDL	BDL
6058	6/15/94	BDL	BDL	BDL	BDL	BDL
6071	6/15/94	1.5	1.0	BDL	BDL	2.5
5271	7/13/94	92	35	75	166	368
5273	7/13/94	144	BDL[a]	BDL	2.2	146
5274	7/13/94	BDL	BDL	BDL	BDL	BDL
6053	7/13/94	BDL[a]	BDL[a]	8.0	12.4	20.4
6054	7/13/94	1.7	BDL	BDL	BDL	BDL
6055	7/13/94	BDL	BDL	BDL	BDL	BDL
6071	7/13/94	2.3	BDL	BDL	BDL	BDL
5271	8/16/94	110	24.2	100	210	444
5275	8/16/94	BDL	BDL	BDL	BDL	BDL
6054	8/16/94	0.99	BDL	BDL	BDL	BDL
6056	8/16/94	0.97	BDL	BDL	BDL	BDL
6060	8/16/94	1.7	BDL	BDL	BDL	BDL
6066	8/16/94	BDL	BDL	BDL	BDL	BDL
6071	8/16/94	BDL	BDL	BDL	BDL	BDL
PS-45	10/4/94	BDL	4	4	39	47
5271	10/6/94	97	30	58	154	339
DL-1	9/11/94	BDL	2.1	BDL	5.5	7.6
ML-3	9/11/94	**6.7**	BDL	BDL	BDL	6.7
ML-5	9/11/94	BDL	1.2	BDL	1.0	1.2
ML-6	9/11/94	BDL	BDL	BDL	2.6	2.6
DL-1	10/4/94	BDL	2.7	1.1	11.8	15.59
ML-3	10/6/94	**6.9**	4.3	1.95	25.5	38.65
ML-4a	10/6/94	BDL	BDL	BDL	BDL	BDL

Note: BDL = below detection level.

[a] BTEX constituents may be obscured by other peaks on GC chromatograph.

1995) contained ethyl benzene at a concentration slightly above the MCL for drinking water, 700 ppb. None of the wells contain toluene or xylene above their respective

Table 14.2 Summary of BTEX Concentrations in Sampled Wells on Various Dates
in 1995

AP Well #	Sampling Date	Benzene (ppb)	Toluene (ppb)	Ethyl Benzene (ppb)	Total Xylene (ppb)	Total BTEX (ppb)
5271	1/20/95	229	40	90	233	613
5782	1/20/95	BDL	BDL	BDL	BDL	BLD
6583	1/20/95	20.0	BDL	28.2	24.8	73.0
6071	1/20/95	2.9	2.7	4.3	BDL	9.9
5271	2/3/95	800	180	327	936	2243
6583	2/11/95	82	BDL	75	51	208
5271	4/12/95	487	145	733	1949	3314
6582	4/12/95	389	11.3	15.0	44	459
6583	4/12/95	72	BDL	79	53	204
6055	4/14/95	BDL	BDL	BDL	BDL	BDL
6063	4/14/95	BDL	BDL	BDL	BDL	BDL
6654	4/14/95	BDL	BDL	BDL	BDL	BDL
5271	6/13/95	84	BDL	32	74	190
6583	6/13/95	123	BDL	100	84	307
ML-3	2/24/95	**10.2**	0.7	BDL	3.5	14.4
ML-3	3/28/95	**7.0**	1.0	BDL	5.7	13.7
ML-5	4/12/95	1.5	0.54	BDL	2.12	4.16
ML-3	4/12/95	**7.8**	0.23	BDL	1.21	9.24
6071	4/12/95	BDL	BDL	BDL	BDL	BDL
MD-10	4/14/95	0.76	0.5	0.27	1.07	2.6
ML-3	6/13/95	6.6	BDL	BDL	BDL	6.6

Note: BDL = below detection level. BTEX constituents may be obscured by other peaks
on GC chromatograph.

MCLs of 1 and 10 ppm. AP6053 appears to be contaminated with an unknown
petroleum product (the sampling crew reported a strong petroleum odor when sampling the well). The chromatographs of the AP6053 samples indicate the presence
of numerous unidentified substances; however, the concentrations were not measured
by the WERC laboratory, as the GC was calibrated to analyze for only BTEX.

The highest BTEX concentrations were found in wells AP5271, AP6582,
and AP6583. BTEX concentrations decreased with distance radiating to the
south, west and east of the above wells (Figure 14.8). However, it should be
noted that four of the wells — AP5782, AP5273, AP5274, and ML-3 — were
sampled when the groundwater levels were above the top of the well screens.
There is evidence that in areas of LNAPL (light nonaqueous phase liquid)
contamination, BTEX concentrations are greater near the water table. Samples
collected when the water level is above the well screen may result in an underestimation of the extent of contamination.

Seasonal variations in analyte concentrations were observed in wells AP5271 and AP6583. That is almost certainly due to the water level dropping close to, or intersecting, the screened interval. Much of the documented contaminants at the Tank Farm are benzene-containing petroleum LNAPLs (Ecology and Environment, Inc., 1994) which tend to float on the water table. LNAPL benzene dissolves and enters the saturated zone closest to the source. The benzene-containing groundwater is then diluted by the process of dispersion and to a lesser extent diffusion as it is carried away from the source. Dilution increases with depth as well as in the lateral direction. The benzene concentrations were greatest when the water table approached the top of the screened interval and then dramatically declined as the water table rose above the screen and the contaminant source moved away from the sampling point.

The concentration gradients found in the bailers and over time after purging at AP5271 and AP7583 indicate that precise quantification of benzene concentrations using the bailer as a sampling device under these conditions is not possible. While the exact cause of the concentration variability is unclear, it is likely due to free or residual product near the water table. When sampling from wells this close to the source of contamination only a qualitative analysis is possible (Light, 1996).

TRACER STUDY

The purpose of this effort was to quantify rates of groundwater velocities and directions within a permafrost-free zone in an area surrounded by permafrost. This exercise also provided critical information on aquifer properties such as dispersivity. As the entire area of Fort Wainwright is such a complex mosaic of frozen/unfrozen ground with seasonally varying groundwater flow rates and directions, no single area presents the ideal location to answer each of these questions with results which may be applied universally to the entire post. However, this test did provide solid confirmation on estimates of dispersivity and velocity.

This site was chosen for the location of the tracer study because it appears that this "unfrozen channel" within the permafrost is an important pathway for movement of groundwater in the area near the Tank Farm. This "channel" in the permafrost appears to be a relic of an ancient slough, which probably caused the initial formation of the thaw bulb under the slough (Figure 14.9). Since the channel is hydraulically connected to other unfrozen areas downgradient, it can serve as a potential pathway for contaminant transport.

Procedure

On August 25, 1994, we injected 875 gal of NaBr with a bromide ion concentration of 806 mg/L over a period of 13.0 h, resulting in an average injection rate of 67.3 gal/h or 4.2 L/min. Immediately after, we injected 500 gal of LiCl over 7.2 h

Figure 14.9 A portion of the "permafrost-free channel" which is a pathway for groundwater movement.

for an average injection rate of 69.0 gal/h or 4.4 L/min. We used a positive displacement pump plus three ball valves to distribute the flow equally to three injection wells. We injected the NaBr into the shallow wells (S1, S2, S3) that were screened from 10 to 20 ft below the groundwater surface (bgw) with 10 slot screens (slot openings of 0.01 in.). The LiCl was injected into a deeper portion of the aquifer with the screen extending from 25 to 35 ft bgw (D1, D2, D3). A descriptive photograph is included as Figure 14.10. The plane view of the injection and monitoring well locations appears in Figure 14.11. The ID of each type of well was 1 in. (2.5 cm). The monitoring wells were screened from 4 to 14, 15 to 25, and 26 to 36 ft bgw for the shallow, middepth, and deep wells, respectively. Monitoring wells were installed at different depths than injection wells to enable detection of changes in vertical flow paths and to quantify flow rate differences between shallow and deeper groundwater. At the time of the August injection, the monitoring wells installed were those extending out to 20 ft hydraulically downgradient from the injection wells. Although we knew the general hydraulic gradient, we proceeded with the injection to gain knowledge about the precise direction and rate of the groundwater flow to help us in planning for subsequent tests.

We then injected a NaBr solution for the second test on September 27, 1994, after the remainder of the monitoring wells (total of 44) were installed. The volume injected this time was 400 gal at a concentration of 672 mg/L of bromide extending over a time period of 11.5 h. Hence, the average injection rate was 34.8 gal/h or 2.2 L/min, about half the rate for the first injection. We decreased the injection rate and the total volume injected for the second test to provide a check as to whether or not the injection rate was having an effect on the groundwater velocity by creating a

Figure 14.10 Photograph of tracer study site, the injection wells and some monitoring wells.

"groundwater mound" in the vicinity of the injection well. We injected this tracer late in the season to get some information on the flow rates when the water was near its peak temperature. The water temperature at the time of injection was about 6°C, which is near the annual maximum. The coldest temperature in the area is about 0°C as water interacts with permafrost.

We both calculated the bromide concentration for the injected solution and measured it with a specific ion electrode. The agreement was good, with the calculated being 6% less than the measured for the first injection and 4% greater for the second injection. A subsequent measured value using ion chromatography was within 1 mg/L of that measured with the specific ion electrode.

After the injection began, we periodically measured ion concentrations (Li^+ or Br^-) for a period of over 1 month at both an injection well and selected monitoring wells. The procedure involved first purging the well of approximately 3 well volumes by extracting groundwater over the screened depth by slowly raising and then lowering a 9.5-mm-diameter PVC tube that was on the suction side of a peristaltic pump. The pumping rate was about 1.0 gpm (3.8 L/min). We then filled a 125-ml PVC bottle with a sample after an initial rinse. These samples were kept at about 4°C for transport back to the laboratory for subsequent analysis using either a specific ion electrode or ion chromatography for the Br and atomic absorption spectrophotometry for the Li. Solutions of 0, 1, 10, 100, 500, and 1000 ppm Br^- were used for calibration of the electrodes.

The remainder of this analysis focuses on use of the bromide data from the second (September 27, 1994) injection to make inferences about the groundwater velocity and dispersion characteristics in the thaw channel. We focused much of our analysis on data collected 30 ft (9 m) downgradient from the injection wells because

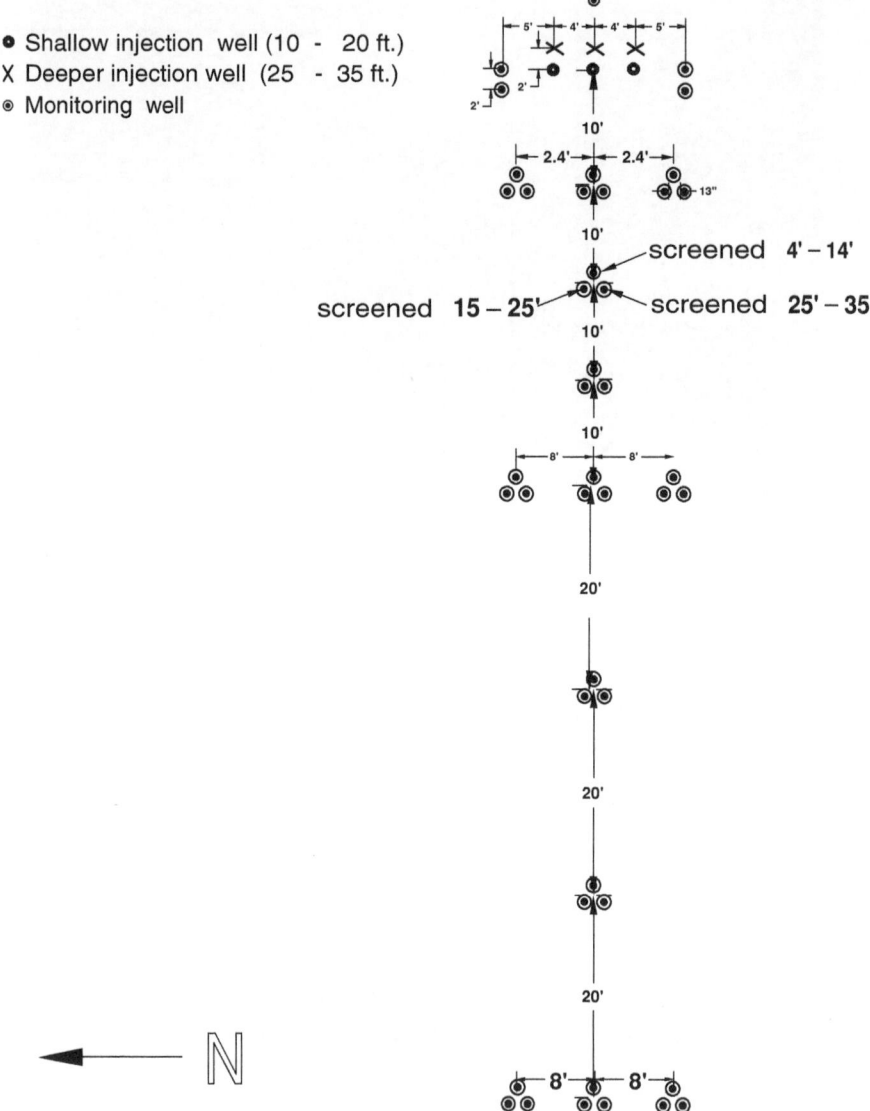

Figure 14.11 Injection and monitoring well layout at tracer study site.

this was both far enough away from the source to be out of the "near field" regime plus the tracer concentrations were high enough to be in a range very amenable to analysis using specific ion electrodes.

Analysis of Tracer Data

We have considered analytical solutions for both one- and two-dimensional transient advection diffusion as one means of utilizing the data obtained from the tracer tests to infer critical parameters such as groundwater pore velocity and longitudinal and/or lateral dispersion coefficients. For the one-dimensional case only, we have considered both a slug injection at one instant in time (t = 0) and a step-function injection with the latter resulting in a Crenel solution.

Let us first discuss the simplest case of one-dimensional transient flow for a slug injection. For an injected mass M in units of mass (kg) per unit area (m²) normal to the flow, the predicted tracer concentration c (ppm) is given by (Bear, 1979):

$$c = \frac{1000M}{\varepsilon\sqrt{4\pi D_x t}} \exp\left[-\frac{(x-ut)^2}{4D_x t}\right] \tag{14.5}$$

Here ε is the porosity, D_x the longitudinal dispersion coefficient (m²/day), t is time (days), x the distance (m) hydraulically downgradient from the injection plane, and u the groundwater pore velocity (m/day). For the second tracer test, the measured concentrations at various times are shown in Figure 14.12 for a longitudinal distance x of 9 m downstream of the injection plane.

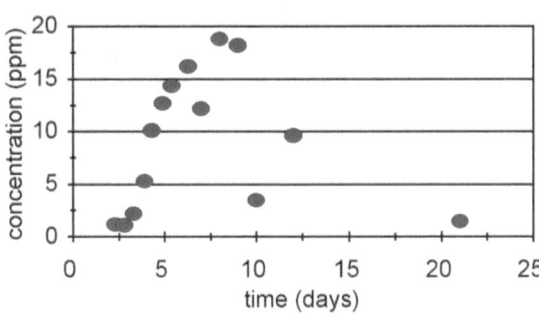

Figure 14.12 Measured bromide concentration at MW 2 for September 27, 1994 injection.

If we now use the solution given by Equation 14.5 for $u = 1$ m/day, $\varepsilon = 0.4$, $D_x = 5.0$ m²/day, and the injected mass per unit area normal to the flow $M = 0.084$ kg/m², we find the results shown in Figure 14.13. We calculated this M from the injected volume of 400 gal at a bromide concentration of 672 mg/L injected into a 3 × 4-m planar area (A_x) normal to the groundwater flow direction. The total injected mass was 1.02 kg.

It is clear that the predicted peak value (about 10 ppm) is lower than that measured by about 60% and occurs slightly sooner (6 days instead of 8 days). In addition, the predicted breakthrough curve rises more quickly and falls off more slowly than the measured.

We next tried a Crenel-type solution (Van Genuchten, 1981), given by Equation 14.6. The differential equation is unchanged from that satisfied by Equation 14.1, but the tracer is now injected as a step function instead of as an instantaneous pulse.

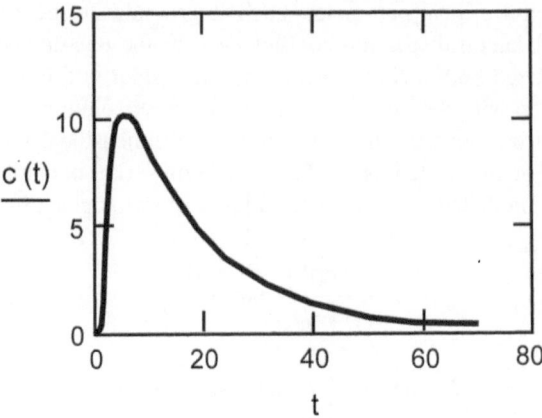

Figure 14.13 Predicted concentration vs. time using one-dimensional model with slug injection.

$$F(t) = \frac{x - u \cdot t}{2\sqrt{D \cdot t}} \qquad\qquad G(t) = \frac{x - u \cdot (t - t_o)}{2\sqrt{D \cdot (t - t_o)}} \qquad (14.6)$$

$$\mathrm{cerf}(z) = 1 - \mathrm{erf}(z) \qquad c(t) = 210 \cdot (\mathrm{cerf}(F(t)) - \mathrm{cerf}(G(t)))$$

Here, D is D_x, t_o is the duration of the F injection (days), $\mathrm{erf}(z)$ is the error function having as argument z, and $\mathrm{cerf}(z)$ is the complementary error function. For the test of interest, $t_o = 0.5$ days and the tracer concentration during the time t_o averaged 421 ppm. The 210 appearing in Equation 14.6 is equal to 0.5 c_o. We calculated c_o by setting the injected mass of 1.02 kg equal to $\varepsilon c_o u A_x t_o$. Using the remaining parameters as before, we find the results in Figure 14.14.

Now, the predicted peak value is quite close to that measured although it occurs somewhat sooner (3 instead of 8 days). The same comments as before apply to the shape of the curve.

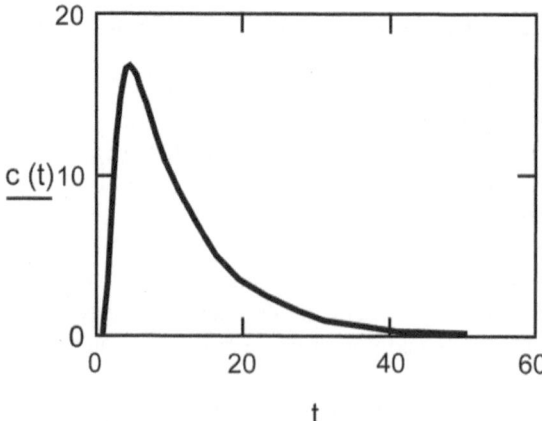

Figure 14.14 Predicted concentration for step function injection.

Last, we tried the solution to the two-dimensional unsteady advection–dispersion equation subject to the instantaneous injection of mass as given by Equation 14.7 (Fried, 1975).

$$c(t) = 1000 \cdot \frac{M}{4 \cdot \pi \cdot \varepsilon \cdot t \cdot \sqrt{D_x(D_y)}} \cdot \exp\left[-\frac{(x - u \cdot t)^2}{4 \cdot D_x \cdot t} - \frac{y^2}{4 \cdot D_y \cdot t}\right] \quad (14.7)$$

Here M is the amount injected per unit depth (3 m) and D_y is the lateral dispersion coefficient. Letting $M = 0.34$ kg/m and $D_y = 0.5$ m²/day leads to the results shown in Figure 14.15. Now the predicted concentrations fall below those measured by roughly a factor of 2 even though the time of predicted peak value is about the same as that measured.

For all three of the predicted curves, the width of the base of the concentration pulse was similar to that measured of around 20 days. This means that the longitudinal dispersion coefficient is not inconsistent with the data. Since $D_x = \alpha_x u$ and we have been letting $u = 1$ m/day, our longitudinal dispersion coefficient is compatible with a longitudinal dispersivity $\alpha_x = 5$ m. This is equal to approximately one half of the scale distance of 9 m to the monitoring wells of interest for this discussion. Such a value is on the high side of what others have found.

To see if the injection rate influenced the downgradient transit velocity for the tracer plume, we compared values for this parameter calculated with data from (1) the first injection and (2) the second injection. We first looked at the times at which the concentration peaked at a given x for up to 40 ft downgradient from the line of injection wells. We discovered that although the larger August injection rate did cause the peak to occur sooner (by a factor of 2) at the closest monitoring wells ($x = 10$ ft or about 3.3 m), for 20 ft, the times to peak only differed by 10% and were essentially the same at the 40-ft locations. Hence, it appears that any disturbance to the natural hydraulic gradient largely disappeared by 40 ft or about 12 m downslope. By this point, the pore velocity as calculated from the travel velocity of the

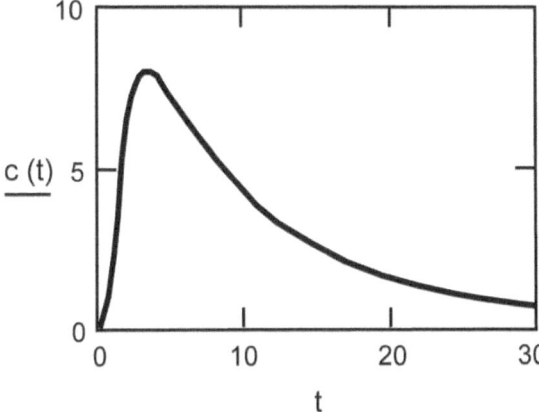

Figure 14.15 Predicted concentration assuming two-dimensional flow and pulse injection.

loci of maximum concentrations was about 1 m/day. These observations are consistent with results from Garabedian et al. (1991), who observed a nonlinear trend in the variance of the locations of the center of mass of a bromide plume during the initial 26 days following injection, which indicated that dispersion was still developing during the early part of their test.

Another intriging result is the maximum concentrations at the 30-ft (9-m) location. The measured peaks were 34 and 18 ppm, respectively, for the first and second injections. One would expect (Equations 14.5, 14.6, or 14.7) that these values would scale linearly with the mass injected. The latter was 2.66 and 1.02 kg of bromide, respectively. The former is calculated using an injected volume of 875 gal at an average bromide concentration of 806 mg/L. The ratio of the peak concentrations at 30 ft was 0.53, while the ratio of the injected masses was 0.38. One reason why the measured peak ratio was greater than the injected mass ratio is that there is some residual bromide from the first test during the second test (our data show values less than 3 ppm, except at the 40-ft location where a value of 12.8 ppm was found 1.5 days after the second injection). This value had to be left over from the first injection because the plume would have to migrate at 27 ft/day if such a high value were a result of the second injection. Another reason is that we did not necessarily collect samples at exactly the time of peak concentration.

Our analyses have assumed a one- or two-dimensional transport process. The data show limited vertical transport over the length (up to 100 ft [30 m]) and time scale (1 month) considered. We injected from 10 to 20 ft below the groundwater surface (bgw) and had the vast majority of the tracer appearing in the middepth monitoring wells that are screened from 15 to 25 ft bgw. For example, for the centerline monitoring wells at 10, 20, and 30 ft, the maximum concentration seen in the shallow or deep wells was about 0.6 ppm, while the middepth wells were found to have concentrations up to 279 ppm, following the first injection. Subsequent to the second injection, the shallow wells along the centerline were found to have concentrations frequently less than 10% of those at middepth. When we look at the overall effect by integrating all the data at each of the three depths, we find that around 3% of the mass is found at the shallow depth, 96% at the midlevel, and 1% at the deep level. So, for our initial analysis, we have focused on the data from the midlevel wells. Summing the ratios of the concentration at the shallow depth to middepth at all locations for which such data existed and then repeating for the deep wells attained these estimates. Then these ratios were weighted by the ratios of middepth concentrations to the maximum middepth concentration and averaged. This allowed us to estimate the total amount of Br and Li tracer in the entire water column at that point downstream from the injection wells.

After we confirmed that analytical Equations 14.5 to 14.7 could reproduce the essential features of the data, we used a more sophisticated analysis technique to infer critical aquifer parameters from the tracer data. As discussed in the literature, methods used to date include that of moments in space (Freyberg, 1986; Garabedian et al., 1991) and curve fitting that minimizes the sum of the square of residuals (Elprince and Day, 1977). The former involves calculating the zeroth moment of concentration over the plume volume to calculate contaminant mass which can be used to infer aquifer porosity. The first spatial moment allows us to calculate the plume center of mass, from which we can infer transport velocity from plotting center of mass location vs. time. The second moment allows us to infer dispersivity values. The method of least squares that we used involved curve fitting the data to Equation 14.5 in such a way that the sum of the squares of differences between the measured values and predicted values was minimized. When we did this for the data shown in Figure 14.12, we found the equation best fit the data for $D \cong 1.8$ m^2/day and $u \cong 0.95$ m/day.

A comparison of this curve fit with the data collected 9 m downstream of the injection wells appears in Figure 14.16. The fit appears rather good except for the one data point at 10 days. The general slopes during the rising and falling portions of the theory match the trend shown by the data quite well. A layering within the aquifer could cause this disagreement between the measured and predicted concentration at 10 days. Such a distribution of heterogeneous materials could cause concentration fluctuations both in space and time due to different hydraulic properties at different depths.

To extend these results to other scales, we must rely on results from the work of others. For example, Gelhar et al. (1992) critically reviewed data relating to field-scale dispersion in aquifers. They found the longitudinal dispersivities ranged from 10^{-2} to 10^4 m over scales from 0.1 to 10^5 m with a variation of up to 3 orders of magnitude at a given scale. Here the dispersion coefficient $D_x = \alpha_x u$ where α_x is the longitudinal dispersivity. Setting the latter to 2 m for our scale of 9 m is consistent with our data. This is toward the high end of the data reported by Gelhar et al. (1992). We would expect the dispersivity to scale with distance linearly. Hence, for making predictions over a scale of a kilometer, we would expect D_x to be no more than 200 m^2/day. The work of others (Gelhar et al., 1992) suggests that it is less than 30% of α_x. Both of these parameters are critical transport variables needed to make contaminant transport predictions.

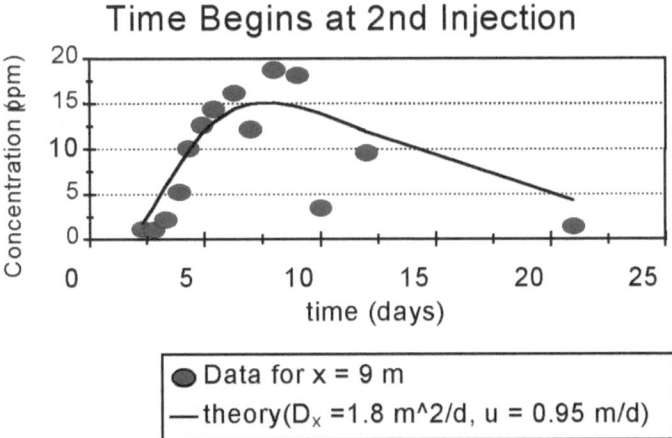

Figure 14.16 Predicted concentration compared with data for September 27, 1994 injection and one-dimensional model given by Equation 14.5.

NUMERICAL MODELING OF CONTAMINANT TRANSPORT
IN DISCONTINUOUS PERMAFROST

Significant masses of discontinuous permafrost have been detected near the source areas during well installations and other activities (Ecology and Environment, Inc., 1994; Lawson et al., 1994). Permafrost, defined as soil or rock where the ground temperature remains below 0°C for 2 or more years (Mueller, 1943), has made predicting the hydrology and contaminant movement extremely difficult and complicated remediation design for the site. Due to these complications, the USACE and CRREL contracted UAF to identify potential contaminant migration pathways and the effect permafrost has on the subsurface hydrology.

With the groundwater elevation and contamination data collected as part of this study and historical data collected as part of other investigations, a computer model was applied to the Tank Farm and TFS source areas. Simulations of the groundwater and contaminant transport processes in the area were completed with SUTRA, a USGS model with single-species solute transport capabilities (Farris, 1996). First, SUTRA was used in conjunction with the data from the tracer study performed within the Tank Farm vicinity in order to get estimates for dispersion coefficients and groundwater velocities. Then the hydrology and contaminant transport within the Tank Farm and TFS area were modeled using benzene as the representative contaminant.

SUTRA

SUTRA, coded with FORTRAN 77, has been in use since 1982, and has been verified with numerous hydrologic projects (Voss, 1984). It simulates unsaturated and/or saturated flow in two dimensions with a constant element thickness. The model uses a unified equation describing single-solute transport or energy transport and is capable of either transient or steady-state simulations for both flow and transport. SUTRA was chosen for this work for numerous reasons. The model is verified and available for public use. The WERC and the USGS worked together closely on this project so support was available in obtaining and is using the model. The code is capable of simulating both energy and contaminant transport. Energy transport in the subsurface was planned for phase two of this project and a model that could accommodate both was needed. Finally, a two-dimensional model was useful for gaining understanding of the subsurface system on a fundamental level with limited data.

Numerical Methods

SUTRA uses the finite element method (FEM) for solving the spatial terms of the differential equations and the integrated finite difference method (IFDM) for temporal terms. The FEM is combined with the Galerkin method of weighted residuals (Voss, 1984). There is an option in SUTRA to use an asymmetric weighting function in order to provide upstream weighting in situations of unsaturated flow where sharp changes in saturation might occur. This option was not used in any of the Tank Farm or tracer simulations since only saturated flow was considered. Under these conditions, the linear basis functions for each node are used as the weighted functions. The Gauss Quadrature method solves the finite element integrals by using weighted sums (Voss, 1984). This method allows exact integration of polynomials by summing point values of the integral (Voss, 1984).

Equations

Unsaturated and saturated flow can be simulated by SUTRA with either steady-state or transient conditions. The hydrology is simulated using the fluid mass balance, presented below:

$$Q_p = \left(S_w \rho S_{op} + \varepsilon \rho \frac{\partial S_w}{\partial p}\right)\frac{\partial p}{\partial t} + \left(\varepsilon S_w \frac{\partial p}{\partial C}\right)\left(\frac{\partial C}{\partial t} - \underline{\nabla} \cdot \left[\left(\frac{kk_r \rho}{\mu}\right) \cdot (\underline{\nabla}p - \rho g)\right]\right) \quad (14.8)$$

where

t	= time
Q_p	= fluid mass source/sink (mass/[length$^3 \cdot$ time])
S_w	= water saturation
ρ	= fluid density (mass/length^3)
S_{op}	= specific pressure storativity ([length $\cdot$ time^2]/mass$_{water}$)
ε	= porosity
p	= pressure (mass/[length $\cdot$ time^2])
C	= solute concentration (mass$_{solute}$/mass$_{water}$)
k	= solid matrix premeability (length^2)
k_r	= relative permeability (length^2)
m	= fluid viscosity (mass/[length $\cdot$ time])
g	= gravity vector (length/time^2)
$\underline{\nabla}$	= $\partial \bar{x} + \partial \bar{y} + \partial \bar{z}$ (1/length)

The left side of this equation represents an external fluid source or sink. The first term on the right side represents the change in the fluid mass due to changes in storativity and saturation, which are a function of the pressure in the subsurface. The second term defines fluid mass changes due to a change in density caused by a solute concentration increase or decrease in the fluid. The third term on the right side can be recognized as a version of Darcy's equation representing fluid mass changes with time due to fluid movement.

If we assume constant density and complete saturation, as has been done for all the simulations discussed in this report, the equation simplifies to:

$$Q_p = \rho S_{op}\frac{\partial p}{\partial t} - \underline{\nabla} \cdot \left[\left(\frac{k\rho}{\mu}\right) \cdot (\underline{\nabla}p - \rho g)\right] \quad (14.9)$$

This equation includes the fluid source or sink term on the left side. The right side has been simplified to only include changes in fluid mass due to storativity variations with time and changes due to differences in the inflow and outflow of the fluid. The density effects with time and saturation changes have been removed by the assumptions.

Mass balances are used to represent solute transport. There are two mass balances to represent both solute mass dissolved in the fluid and solute mass adsorbed onto the soil grains. They include terms for mass changes due to fluid flow (advection), dispersion, diffusion, flux between adsorbed and dissolved components, biological production or decay, and sources or sinks. The following is the equation for solute mass in solution:

$$\frac{\partial(\varepsilon S_w \rho C)}{\partial t} = -\underline{\nabla} \cdot (S_w \rho \underline{v} C) + \underline{\nabla} \cdot [\varepsilon S_w \rho (\underline{D} + D_m \underline{I}) \cdot \underline{\nabla} C] \qquad (14.10)$$
$$- f + \varepsilon S_w \rho \Gamma_w + Q_p C^*$$

where

$\underline{v}$ = average pore velocity (length/t)
$\underline{D}$ = dispersion tensor (length²/t)
D_m= molecular diffusivity (length²/t)
$\underline{I}$ = identity matrix
f = volumetric adsorbate source (solute adsorbtion rate)
        [mass$_s$/ (length³ × t)]
G_w= solute mass source in fluid due to production reactions
        [mass$_s$/ (mass × t)]
C^*= solute concentration of source/sink (mass$_s$/mass)

The mass balance equation for adsorbed solute mass has similar terms and is shown below:

$$\frac{\partial[(1-\varepsilon)\rho_s C_s]}{\partial t} = f + (1-\varepsilon)\rho_s \Gamma_s \qquad (14.11)$$

where

ρ_s = density of solid grains (mass$_G$/[length$_G^3$])
Γ_s = solute mass source of adsorbed solute due to production reactions
        (mass$_s$/[mass × t]
C_s = concentration of solute on soil grains (mass$_s$/mass$_G$)

The left-hand side of both solute balances simply represents the change in mass with time. On the right-hand side of the dissolved mass balance, the first term represents the change in mass due to advective differences and the second term represents mass change due to dispersion and diffusion. These terms are not included in the sorbate balance since advection and diffusion do not apply to stationary soil grains. The f term that appears in both equations is a partitioning variable represent-ing the solute adsorption rate. A biological production term also occurs in both equations to account for production or decay of solute mass due to biological activity in the fluid or on the grains. Finally, in the dissolved solute balance, there is a source/sink term. Also in the dissolved solute equation, if we assume saturation, S_w becomes one and drops out of the equation as in the fluid mass balance.

User-Supplied Data

The user must specify boundary conditions, grid size, spatial and temporal discretization, hydrologic parameters, and either contaminant or thermal properties. This study did not include thermal properties. Therefore only the contaminant transport aspect will be discussed.

Three types of boundary conditions can be specified in SUTRA: specified pressure or head at a node, specified concentration at a node that has a specified pressure, and an inflow or outflow with or without a specified concentration.

The grid size and spatial discretization are specified in the mesh generator program which is a preliminary processor that creates the elements and numbers the nodes to minimize the bandwidth. This information is input directly to the SUTRA input file D5. The grid size is dependent to some degree on the required discretization and the capability of the computer. General rules-of-thumb presented in the SUTRA manual (Voss, 1984) for spatial discretization are:

$$L_x < 4 \times \alpha_L \text{ and } L_y < 10 \times \alpha_T \qquad (14.12)$$

where:

L_x = length of an element in the x-direction (length)
L_y = length of an element in the y-direction (length)
α_L = longitudinal dispersivity (length)
α_T = transverse dispersivity (length)

This first criterion is based on a unitless mesh Péclet number defined below (Voss, 1984):

$$Pe_m = \frac{\varepsilon \underline{v} \Delta L_L}{|\varepsilon(\sigma_w + \alpha_L \underline{v}) + (1 - \varepsilon)\sigma_s|} \qquad (14.13)$$

where

σ_w = diffusion in the fluid
σ_s = diffusion in solid phase
L_L = distance between element sites

If the diffusion can be ignored, the above equation simplifies to distance between elements over longitudinal dispersivity. Voss (1984) states that spatial stability is usually achieved with SUTRA if the Peclet number is less than 4 and, thus, the 4 in Equation 14.12. The second criterion has no exact mathematical basis, but is presented by Voss (1984) as a general rule of thumb that provides a small transverse length compared to the transverse dispersion. If the stability is not certain (i.e., negative concentration, unusual flow patterns), an incremental increase and/or decrease in the discretization can be attempted to determine the effect of the discretization on the results.

Tables 14.3 and 14.4 summarize the required hydrologic and contaminant parameters that need to be specified.

Output options in SUTRA include the following:

- Concentrations at every node
- Mass budgets — fluid and/or solute
- Velocities at each node
- Pressures or heads at each node
- Degree of saturation
- Observations — concentrations at specific nodes
- Restart data

TRACER STUDY SIMULATIONS

The purpose of the tracer study was to quantify groundwater velocities and directions, as well as aquifer properties, such as dispersivities, in a permafrost-free

Table 14.3 Hydrologic Parameters in SUTRA

Parameter	Symbol	Dimensions
Hydrologic boundary conditions	p or q	(mass)/(length $\times$ t^2) or mass/t
Fluid compressibility	β	(length $\times$ t^2)/mass
Fluid diffusivity	σ_w	length^2/t
Fluid density	ρ	mass/length^3
Fluid viscosity	μ	mass/(length $\times$ t)
Soil compressibility	α	(length $\times$ t^2)/mass
Soil density	ρ_s	mass/length^3
Porosity (per node)	ε	unitless
Aquifer thickness (per node)	t	length
Permeability or hydraulic conductivity (per element)	k or K	length^2 or length/t

Table 14.4 Contaminant Transport Parameters in SUTRA

Parameter	Symbol	Dimensions
Concentration boundary conditions	C_o or C^*	mass$_s$/mass
Dispersivities (per element)	σ_L and σ_t	length
Adsorption type	Linear, Freundlich, Langmuir	not applicable
Adsorption distribution coefficient	χ_1	length$_f^3$/mass$_G$
Freundlich or Langmuir coefficient	χ_2	length or length$_f^3$/mass$_s$
Zero or first-order rate of production of solute in fluid	γ_0^w or γ_1^w	mass$_s$/mass/t or t^{-1}
Zero or first-order rate of production of solute in soil	γ_0^s or γ_1^s	mass$_s$/mass$_G$/t or t^{-1}

zone surrounded by impermeable boundaries of permafrost, extending from near the surface to bedrock at depth. This tracer study site was chosen because it is suspected that the unfrozen "channel" within the permafrost is an important pathway for groundwater movement and contaminant transport.

Simulation Development

With these parameter estimates from the analytical solutions, SUTRA was used to simulate the tracer study using a cross-sectional grid (Figure 14.17). The cross-sectional grid was set such that the constant dimension in the simulation would be the width of the unfrozen channel normal to the flow (the y-direction) as this dimension was constrained by the permafrost. The data collected indicated that over 90% of the tracer mass was contained within the well configuration, which was 18 ft wide and 100 ft long. The grid dimensions, therefore, were $100 \times 18 \times 50$ ft with the elements being discretized to $1.0 \times 18.0 \times 1.0$ ft.

The simulation was broken into two parts, the 12-h injection period and the 31-day monitoring period, using the hot start option for the latter. For both periods, steady-state flow with transient contaminant transport was utilized. The injection period was simulated with 7.5-min time steps, while the 31-day monitoring period was simulated with 3-h time steps. The 12-hour period simulated the injection of 400 gal of tracer over 12 h into 11 nodes that represented the 10 ft of screen on the injection wells. These conditions resulted in an input water flow of 3.18 cm^3/s (1.12×10^{-4} ft^3/s) per node. The input concentration was 672 mg/L.

Since constant density for water was assumed and the density of the tracer solution and the water were essentially equivalent, gravity effects were neglected and the hydraulic gradient was simulated using hydraulic heads at both the east and west ends of the grid. During the study, head values at the injection point, the eastern boundary of the grid, were measured using a pressure transducer that was placed in an injection well immediately following completion of the tracer injection. A head value at the western end of the grid was calculated using a measured hydraulic

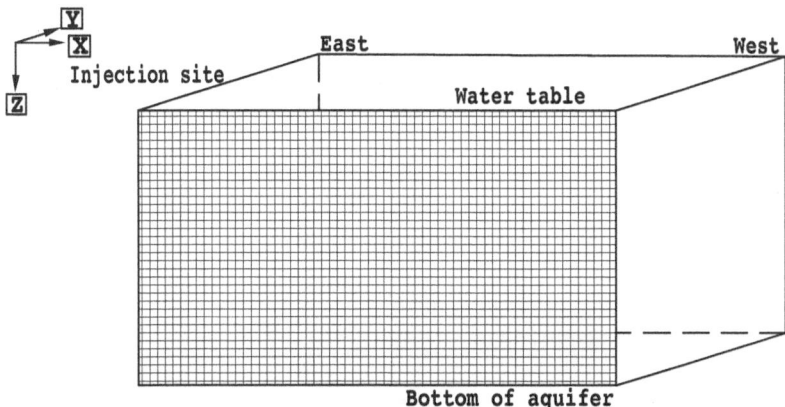

Figure 14.17 Grid for the tracer simulation.

Table 14.5   Parameter Values for Tracer Simulation

Parameter	Value
Fluid compressibility	0
Longitudinal dispersivity	2.0 m (6.6 ft)
Transverse dispersivity	0.2 m (0.66 ft)
Fluid density	1 (unitless)
Fluid viscosity	1 (unitless)
Soil compressibility	0
Soil density	2650 kg/m^3 (165 lb/ft^3)
Adsorption type	none
Adsorption distribution coefficient	0
Freundlich or Langmuir coefficient	0
Zero or first-order rate of production of solute in fluid	0
Zero or first-order rate of production of solute in soil	0
Grid size (x, y, z)	30.5 m, 6 m, 15.2 m
	(100.0 ft, 19.7 ft, 50.0 ft)
Element size	0.3 m, 6 m, 0.3 m
	(1.0 ft, 19.7 ft, 1.0 ft)
Injection input per node	3.19 × 10^{-6} m^3/s
	(1.12 × 10^{-4} ft^3/s)
Injection concentration	672 mg/L
Pressure head at injection point	130.028 m (426.60 ft)
Pressure head 100 ft downgradient from injection point	129.94 m (426.31 ft)
Pore velocity	1 m/day (3.3 ft/day)
Hydraulic conductivity	1.5 × 10^{-3} m/s (0.0049 ft/s)
Porosity	0.4

gradient of the area of 0.003 (Ecology and Environment, Inc., 1994) and the value at the injection point. The actual values are listed in Table 14.5 along with other parameter values.

The hydraulic conductivity was calculated using Darcy's equation with a pore velocity of 1 m/day, porosity of 0.4, and a hydraulic gradient of 0.003. Tables of typical values for sands and gravels in Freeze and Cherry (1979) were used to estimate the porosity. The dispersion values were obtained from the analytical regression discussed earlier and converted to dispersivities for the model using the following equation (Voss, 1984):

$$\alpha_L = \frac{(D - D_m)}{v} \tag{14.14}$$

where

α_L = longitudinal dispersivity (length)
D = longitudinal dispersion coefficient (length/t)
D_m = molecular diffusivity (length^2/t)
v = pore velocity (length/)

Diffusion was assumed to be insignificant due to the velocity of the groundwater. Isotropy was assumed for numerical simplification and the lack of data to justify more complexity.

Results

Figure 14.18 shows the results of the simulation with these initial conditions at three separate time steps. The first time step shows the tracer plume at the end of the injection period. The tracer has not moved more than 15 ft downgradient and the contours are close together, indicating steep gradients in the concentration near the injection nodes. At 15 days after injection, the highest concentration of tracer is observed traveling along a flow path from east to west

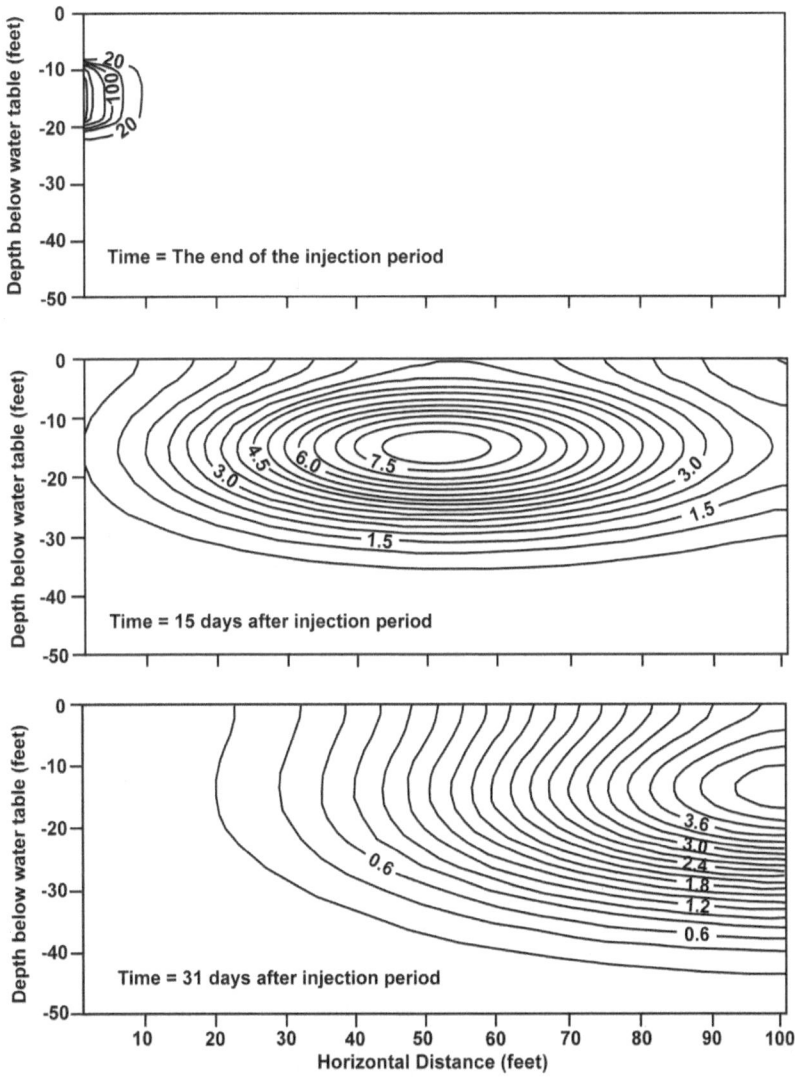

Figure 14.18 Tracer simulation bromide concentration contours (mg/L).

at approximately 10 to 20 ft below ground surface (bgs), the depth of injection. After 30 days, the tracer progresses beyond the grid at concentrations of approximately 4 mg/L. Longitudinal dispersion is dominant with the plume stretching over 300 ft along the x-axis and only 100 ft along the z-axis. The data indicating that over 90% of the tracer was within 15 to 25 ft below the water table corroborate this.

These results were compared with the measured field data by means of the root mean square error (RMS). The RMS equation is defined below:

$$RMS = \left[\frac{1}{n} \sum_{i=1}^{n} (h_m - h_s)_i^2 \right]^{1/2}$$ (14.15)

where

n = number of values
h_m = measured value
h_s = SUTRA result

The RMS value for this simulation was 5.25 using field data from all of the centerline wells at 15 ft below the water table. This subset of data was used as it was the most complete and most consistent. The optimum set of parameter values may be determined through an iterative process that would achieve the lowest RMS value.

With this value as a starting condition, a sensitivity analysis was conducted to determine which parameters most affect the RMS value and the model results. The analysis incorporated porosity, hydraulic conductivity, longitudinal dispersivity, and grid thickness, as there was some degree of uncertainty with each of these parameters and they had the potential to affect the model results significantly. Hydraulic gradient was also included in the analysis to determine if the seasonal or annual fluctuation of that parameter would affect the tracer results.

The analysis was conducted by using the initial parameters that were justified by either the field measurements or the literature. One of the five parameters was changed by –50%, –20%, +20%, and +50%, while the other four parameters remained constant at the initial conditions. RMS values were calculated for each set of results as well as the percent change from the initial simulation RMS value. Figure 14.19 shows a graph of the percent change in the RMS values vs. the percent change of the parameters.

The greatest change in the RMS values occurs when changes are made to the grid thickness and to soil porosity. The model results are not as sensitive to the hydraulic conductivity, longitudinal dispersivity, and hydraulic gradient.

The absolute RMS values from the sensitivity analysis showed that the initial simulation does not have a "best-fit" set of parameters. The initial sensitivity was, therefore, extended to explore a better set of parameters. Three of the parameters were changed beyond +50% in the direction that lowered the RMS values. The dispersivity parameter was changed +70%, +80%, +90%, +100%, and +150% in order to reach a minimum RMS value. The thickness and the hydraulic gradient

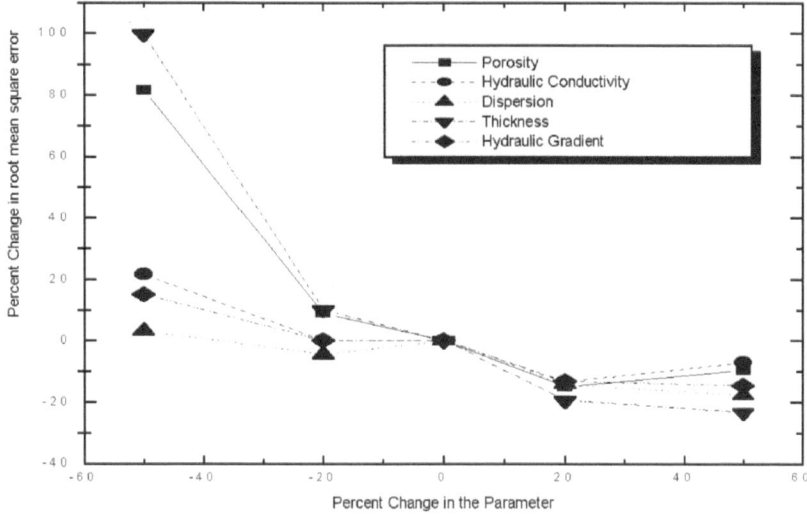

Figure 14.19 Sensitivity analysis for the tracer simulation.

were changed +70% to reach a minimum RMS. Porosity and hydraulic conductivity had already reached a minimum at +20% change.

After a minimum RMS value was achieved for each of the single parameter changes, several combinations of these "best" values were attempted. The best-fit parameters achieving the lowest overall RMS values of 3.9 are listed in Table 14.6.

Table 14.6   Best-Fit Parameters for Tracer Simulation

Parameter	Best-Fit Value
Porosity	0.4
Hydraulic conductivity	1.8×10^{-3} m/s
Longitudinal dispersivity	2.0 m
Grid thickness	3.75 m
Hydraulic gradient	0.003

Discussion

Figure 14.20 shows breakthrough curves that compare the results from the best-fit SUTRA simulation with the measured field data downgradient of the injection wells. This curve represents the tracer breakthrough along the centerline of the tracer plume. Depth below the water table was 15 ft (4.6 m). The one-dimensional analytical results were also included on the curve to show another comparison to the field data. The analytical results predict a peak concentration of 15 mg/L occurring between 7 and 8 days after injection. This matches the peak time recorded in the field, but the concentration is approximately 4 mg/L less than was measured. SUTRA predicts a peak concentration of 8 mg/L at 6.5 days. The concentration is 11 mg/L lower than the measured data and occurs a

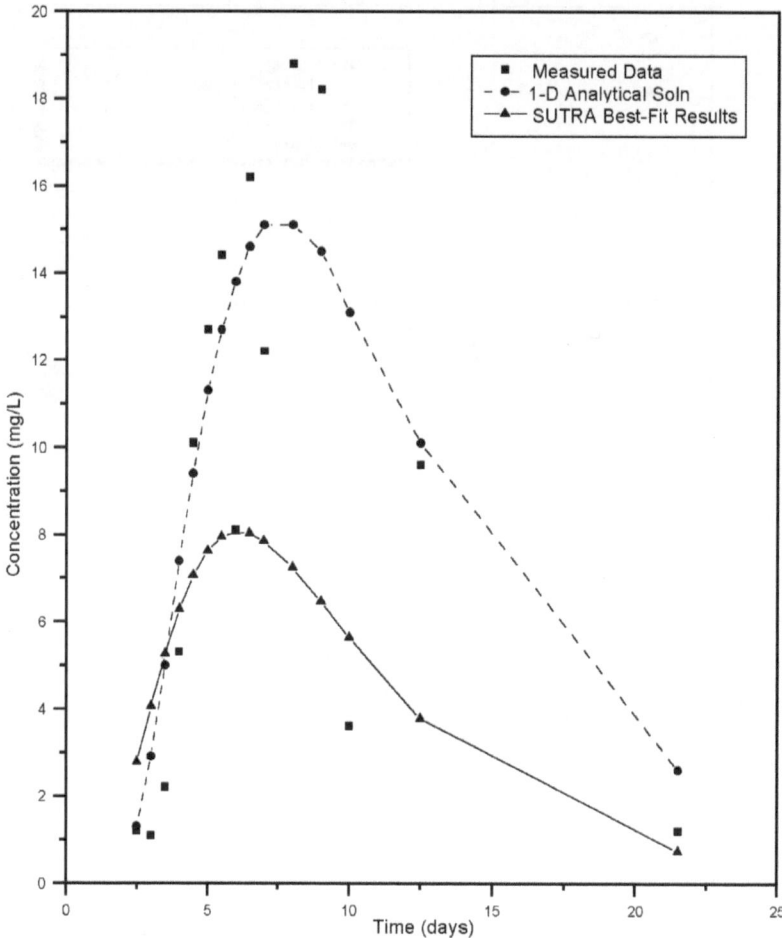

Figure 14.20 Tracer breakthrough curve at a distance of 30 ft downgradient.

day earlier. The measured concentrations of the tracer have a minimum error of ±1 mg/L.

The one-dimensional results appear to give a better fit than the two-dimensional SUTRA results, which may seem counterintuitive. This discrepancy can be explained, however, by looking at the quality of the field data and the effects of the third dimension in the SUTRA model. The one-dimensional analytical results were obtained by utilizing the data measured 30 ft downgradient, whereas the SUTRA results were obtained by calibrating the model to all the field data collected. The data from the other locations, particularly 60, 80, and 100 ft, showed more hetero-geneities and anomalous readings than the data at 30 ft downgradient. These anom-alous measurements may have skewed the SUTRA results.

The third dimension in SUTRA, the grid thickness, also skews the concentration results. The concentrations in SUTRA are averaged over the constant third dimension, which serves to generally lower the peak concentrations and increase the tail values. If the tracer simulation was run with a grid thickness equal to zero (as is the case in the one-dimensional solution) and the mass injection was adjusted accordingly, the SUTRA results would match the one-dimensional analytical results more closely. This was not done for this work because the objective of using SUTRA was to consider all the field data and dispersion in each direction. A comparison of mass input and mass output for the model was calculated. The mass input was calculated by multiplying the injection flow at each of the 11 nodes of 3.18×10^{-6} m^3/s (1.12×10^{-4} ft^3/s) by the input concentration of 672 mg/L and by the injection period of 12 h. This gave a total mass input of 1.02 kg (2.2 lb). The output was calculated by multiplying the pore volume of one element by the sum of the concentrations at each node in the grid. This is the equivalent of multiplying the total concentration by the water volume in the grid. The total output for the best-fit simulation is 1.06 kg (2.3 lb). The relative percent difference is 4%, indicating a small amount of mass being produced.

The low peak concentrations may be partly explained by the lateral integration of the tracer in the third dimension. The tracer concentration did not vary in the third dimension due to the nature of SUTRA, so the concentration was averaged out in that direction. This averaging served to lower the peak values. This would explain why the largest underestimates by SUTRA occurred at 30 and 40 ft downgradient where the highest field concentrations were recorded.

It is quite possible there are significant heterogeneities in the subsurface that were not included in the simulations. Areas of low permeability or porosity within the channel due to segregated soils would cause the flow to change course within the channel. It was clear from these results, however, that dispersion, particularly longitudinal, was a crucial process in the tracer transport and that the permafrost-free channel was a hydrologically active area. The flow was from east to west in this channel and the pore velocity was approximately 1 m/day.

Tank Farm Simulations

The Tank Farm simulation was developed in order to meet the objective of utilizing a computer model to represent the areawide hydrology as well as determine the effects discontinuous permafrost would have on contaminant transport. The simulation was run with and without incorporating the permafrost configuration determined by Lawson et al. (1994). Both approaches used the same boundary conditions, grid size and element discretization, and general aquifer properties. The results of the two simulations were compared to determine the overall effect the permafrost has on the hydrology and transport. The simulation results were compared with the field measurements of BTEX concentrations as a reference for the location and relative magnitudes of the contaminant. The measurements were taken in wells with varying screened intervals and thus represent concentrations at various depths in the aquifer (Light, 1996). SUTRA results represent an average concentration over the thickness of the elements, so they do not correspond with the smaller screened intervals of the wells. The numbers could not, therefore, be compared directly.

General chemical parameters measured in the wells were used to help understand the physical, chemical, and biological processes occurring in the aquifer. A charge balance on the ions in solution indicates the water is generally well characterized by the measured ions. The hardness (the sum of calcium and magnesium concentrations) and alkalinity values also agree closely, supporting this conclusion.

The ions and alkalinity were used to estimate the major source of the aquifer recharge, utilizing a comparison method described in Lilly and Ray (1987). Water flowing from the Chena River has a different ionic composition than rain or snow infiltration. Therefore, it was possible to get an indication of the source of recharge for the groundwater in the Tank Farm area by comparing the ionic composition of the water samples taken in monitoring wells within that area with samples taken in monitoring wells near the river. An analysis of alkalinity concentration — as calcium carbonate and potassium concentration from monitoring points at the river, between the river and Birch Hill, and near Birch Hill — demonstrates that the area near the Tank Farm is hydrologically distinct from the area near the river. The samples from wells near the TFS and Tank Farm show a closer resemblance to the Birch Hill well chemistry than to the river chemistry. This agrees with the hydraulic heads in the area, indicating flow to the southwest.

Grid Development

The SUTRA grid designed for the Tank Farm area (Figure 14.21) utilized natural boundary conditions including the Chena River, Noyes Slough, and the bedrock that defines Birch Hill. The grid dimensions were 7000 × 7000 ft horizontally × 100 ft vertically with individual elements being 100 × 100 × 100 ft.

The east, south, and west edges of the grid used specified head boundary conditions determined by gathering all the water levels taken in the wells in the Tank Farm area, on the Chena River, and on the Noyes Slough for the years 1994 and 1995 and calculating an arithmetic average for each measuring site. This time period contained the most extensive data and coincided with the contaminant measurements that were collected. The average head values were contoured using the Kriging method in the SURFER® software (Figure 14.22). The Kriging method preserved the original data points while interpolating between the known values, thus the measured values were not compromised (Keckler, 1995). Contour lines were placed over the SUTRA grid and heads were correlated to specific nodes on the east, south, and west boundaries. Linear interpolation was used for nodes falling between contour lines.

The northern edge of the grid was defined by Birch Hill. The nodes that covered the hill area were considered to be no-flow boundaries, while the other nodes were defined with a specified flux boundary based on recharge values to be discussed later. The no-flow assumption for Birch Hill was based on work done by the USGS (Bolton, 1996). A simulation of the bedrock hills, including Birch Hill, indicated flow rates from the hill to be orders of magnitude less in most areas than those in the floodplain. The simulation also showed little upward vertical flow, which supported the assumption that the bottom of the grid is a no-flow boundary.

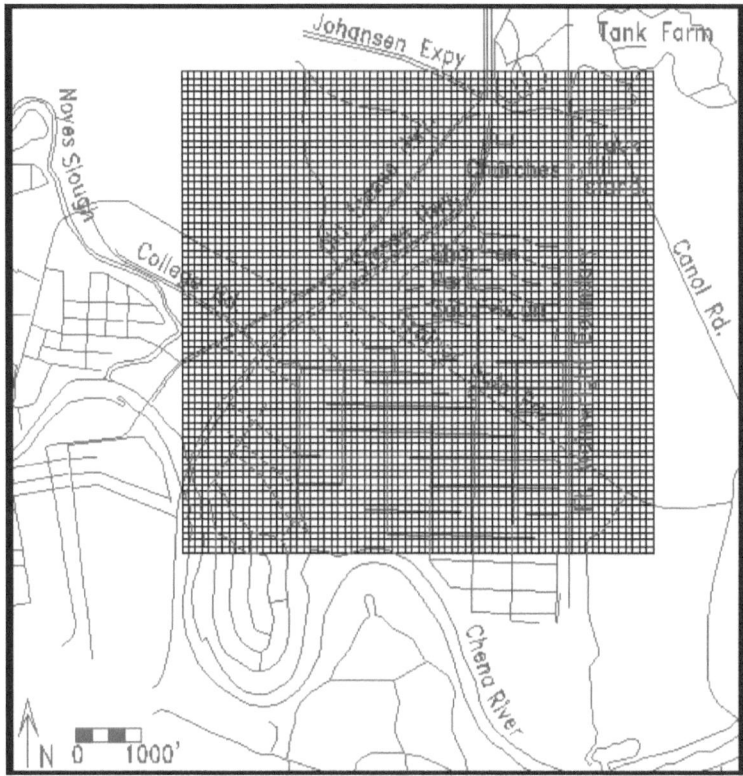

Figure 14.21 SUTRA grid for Tank Farm simulation.

The approximate average depth to bedrock in the area is 100 ft. This is not an accurate depth on Birch Hill which consists mostly of bedrock, but further from the hill, in the main area of interest, this is a more reasonable value and suitable for an initial value.

Hydrology

We first simulated the subsurface hydrology of the Tank Farm area, ignoring the contaminant transport. The permafrost configuration was a critical component of the hydrology simulations. Simulations were completed with and without permafrost in order to investigate the effects of permafrost on the groundwater flow patterns.

Permafrost

Permafrost was included in the model by adjusting the hydraulic conductivity for nodes with varying degrees of permafrost. First, the permafrost configuration illustrated by Lawson et al. (1994) was placed over the Tank Farm grid and discretized into the grid elements (Figure 14.23). The four shades of gray in Figure 14.23 represent different amounts of permafrost in the ground profile applied to the

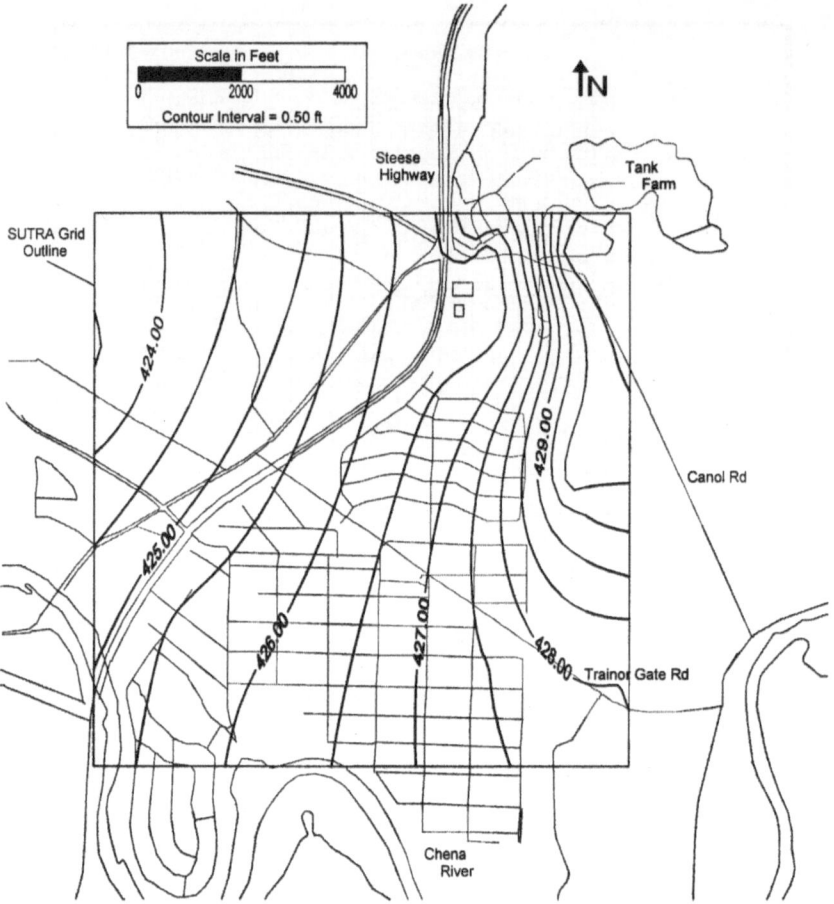

Figure 14.22 Average measured head values (ft) in Tank Farm vicinity.

grid elements. Elements with no shading are permafrost free. These scenarios were derived from combining layouts of the top and bottom surface of the permafrost configuration shown earlier in Figures 14.3A and 14.3B. In discretizing these figures, some small-scale anomalies and variations were eliminated that should be noted as a limitation of the element size.

The first possible element scenario was no permafrost (the white elements in Figure 14.23), and this was given a hydraulic conductivity of 1.8×10^{-3} m/s (508 ft/day), which was the best-fit value for conductivity in the tracer model. The second scenario was a permafrost "wall," frozen soil for the entire 30 m (100 ft) of aquifer (the black elements). The elements having this scheme were given a hydraulic conductivity of 1×10^{-6} m/s (0.28 ft/day) based on the work of Kane and Stein (1983b), which concluded that frozen soil could have a hydraulic conductivity several orders of magnitude less than the same soil type unfrozen.

The final three scenarios fall within these two extremes. The third configuration was unfrozen soil for the first few meters, but frozen for the rest of the depth (lightest

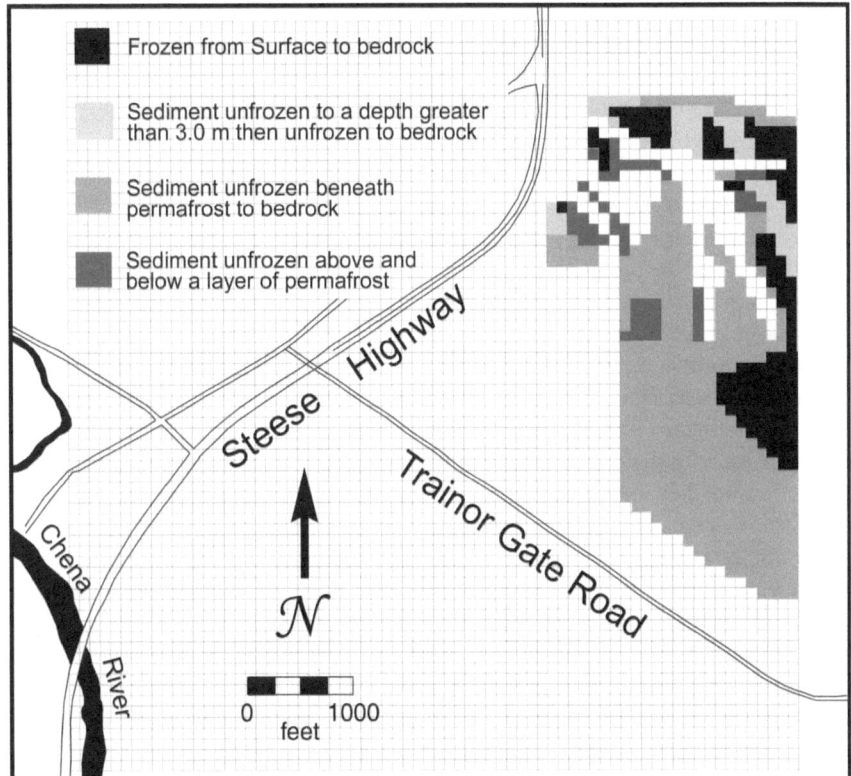

Figure 14.23 Discretized permafrost configuration for Tank Farm grid.

gray elements). An average unfrozen thickness was assumed to be 5 m (16.4 ft) because Lawson et al. (1994) listed this area as being unfrozen for several meters bgs and there are wells in that area that extend 6 to 9 m (20 to 30 ft) bgs that are unfrozen. Water depth in these wells was typically 3 to 4 m (10 to 13 ft) bgs so the saturated thickness would be approximately 5 m (16.4 ft). The hydraulic conductivity for this condition was assessed by the following formula:

$$K = R_u K_u + R_f K_f \tag{14.16}$$

where

K = element hydraulic conductivity
R_u = ratio of unfrozen depth to aquifer depth
K_u = unfrozen soil hydraulic to conductivity, 1.8×10^{-3} m/s
R_f = ratio of frozen depth to aquifer depth
K_f = frozen soil hydraulic conductivity, 1.0×10^{-6} m/s

This equation gave a hydraulic conductivity of 3.008×10^{-4} m/s (85.3 ft/day) for the third scenario. The fourth configuration was frozen near the ground surface and unfrozen for the rest of the depth (medium gray elements). An unfrozen thickness of 10 m (32.8 ft) was assumed as a guess based on the Lawson et al. (1994) GPR work and the aquifer thickness of 30 m. The above equation was used to determine a hydraulic conductivity of 6.007×10^{-4} m/s (170 ft/day). The final scenario occurred when a frozen layer existed between two unfrozen soil sections (darkest gray elements). For this condition, the assumed thickness was 15 m (49.2 ft) (a combination of the third and fourth scenarios) and the hydraulic conductivity was 9.005×10^{-4} m/s (255 ft/day).

Recharge

Recharge was calculated using average precipitation of 10.84 in/year (27.53 cm/year), recorded at the Fairbanks International Airport from 1961 to 1990 (National Climate Data Center, 1995) and an estimate of ET from Dingman (1971) of 6 in/year (15.24 cm/year). The recharge value for each node was first calculated assuming no permafrost, thus, no runoff and constant storage. The recharge could then be determined by subtracting ET from precipitation which gave 4.84 in/year (12.29 cm/year). The area of the grid was 4.9×10^7 ft^2 $(4.55 \times 10^6$ m$^2)$ with 5041 nodes in the grid. Multiplying the recharge by the grid area and dividing by the number of nodes gave a recharge-per-node value of 3.52×10^{-6} m^3/s $(1.24 \times 10^{-4}$ ft/s).

In the area where there was no permafrost or permafrost started below the water table (permafrost scenarios 1, 3, and 5), runoff was assumed to be zero and the above recharge value was used directly. On Birch Hill and in areas where permafrost started above the water table (permafrost scenarios 2 and 4), recharge was assumed to be zero. The runoff from these areas was introduced at the nearest "open" node, out of bedrock or unfrozen, in the direction of the river (usually south). If the nearest open node was further than 400 ft away, the runoff was assumed to be depleted by ET.

Hydrologic Simulation Setup

With the above grid setup, the steady-state subsurface hydrology in the Tank Farm area was simulated before transport was introduced. The hydrology was simulated with and without the permafrost configuration to determine the effects the permafrost has on the flow paths, as these will affect the transport pathways. The results of each of these simulations were compared with data from the specific observation sites as well as the general contoured picture of the average water elevations. The hydraulic gradient over the area was 0.0002, measured from the averaged water elevation measurements. This gradient indicated that the elevations varied slightly within the Tank Farm area and came within the accuracy of the electronic measuring device used to record the water elevation. Thus, although the error might have been small, the results were not necessarily representing accurate flow. The overall contour map was, therefore, important for a qualitative comparison of the flow field to determine if the SUTRA pathways matched the pathways indicated by the measured data. The general parameters used for both hydrologic simulations are listed in Table 14.7. The transport parameters were not applicable for these simulations. The hydraulic conductivity and porosity were for unfrozen soil only and resulted from the tracer study

Table 14.7   Parameter Values for Tank Farm Hydrologic Simulation

Parameter	Value
Fluid compressibility	4.4×10^{-10} m²/N
Longitudinal dispersivity	N/A
Transverse dispersivity	N/A
Fluid density	1000 kg/m³
Fluid viscosity	1.0×10^{-3} kg/(m s)
Soil compressibility	10^{-8} m²/N
Soil density	2650 kg/m³
Adsorption type	N/A
Adsorption distribution coefficient	N/A
Freundlich or Langmuir coefficient	N/A
Zero or first-order rate of production of solute in fluid	N/A
Zero or first-order rate of production of solute in soil	N/A
Grid size (x, y, z)	$2135 \times 2135 \times 30$ m
Element size	$30 \times 30 \times 30$ m
Injection input per node	N/A
Injection concentration	N/A
Hydraulic conductivity for unfrozen soil	1.8×10^{-3} m/s
Porosity	0.4

best-fit parameters. The fluid and soil compressibility were taken from tables in Freeze and Cherry (1979) for water and a silty sand. The soil density was also taken from Freeze and Cherry (1979) for a silty sand.

Hydrologic Results

The results for the two separate simulations (shown in Figures 14.24A and 14.24B) show a marked difference in the northeast corner of the grid, which corresponds to the location in the Tank Farm and TFS areas where the majority of permafrost is found. In Figure 14.24A, with no permafrost, the contours are smooth and slope gently to the southwest toward the river. Figure 14.24B, with the permafrost, shows sharper contours and bending lines in the northwest corner. These are the same patterns, although not to the same extent, seen in the measured elevation contour plot. The permafrost seems to cause a mounding of the water table to the east, forcing the water around the large, frozen zones. The differences between the two simulated results fade closer to the river and the slough to the southwest. The anomalies on the eastern boundary in the simulation without permafrost result from the boundary conditions. The measured water levels used for the boundary conditions in both simulations were in a permafrost area, so these effects were not eliminated in the nonpermafrost simulation.

The simulation with permafrost shows a closer resemblance to the measured groundwater levels, but the simulated contours do not show all the features of the measured elevations. A RMS error was calculated between the measured average elevations and two sets of simulated results using the RMS error equation described previously. The error between the measured values and the simulation without permafrost was 0.287, while the error associated with the simulation including permafrost was 0.152. A sensitivity analysis was conducted to explore the differences

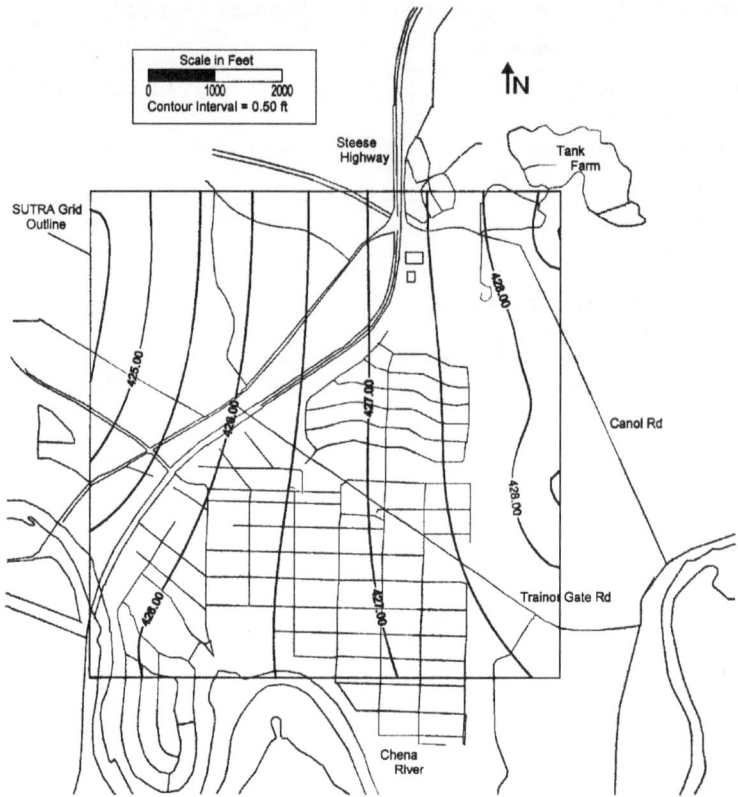

Figure 14.24A Water elevation contours for Tank Farm simulation without permafrost.

further. The varied parameters included hydraulic conductivity for the frozen soil (K_f), hydraulic conductivity for unfrozen soil (K_u), porosity, and recharge. Each parameter was changed −50%, −20%, +20%, and +50%, while the other three parameters remained at the initial conditions described above. The results were compared using the RMS error calculated between the average measured water elevations and the simulated elevations.

The sensitivity analysis showed that the porosity and K_f had little effect on the simulation results. The recharge and K_u almost equally affected the results. These results indicate that if recharge was taken to be zero and K_u was increased further, the error would decrease further. A low RMS error of 0.141 was reached at K_u of 2.70×10^{-3} m/s (8.86×10^{-3} ft/s). A further increase in K_u increased the error. A minimum error value was never confirmed with the recharge parameter, as the error continued to decrease with decreasing recharge. Decreasing the recharge further would likely give a lower RMS error, but would not be within the range of realistic values. The values considered to be the most reasonable, those that result in a low RMS of 0.141, are presented in Table 14.8.

There are several possible reasons for why the simulated results do not match the measured values exactly. First, the permafrost configuration significantly affects

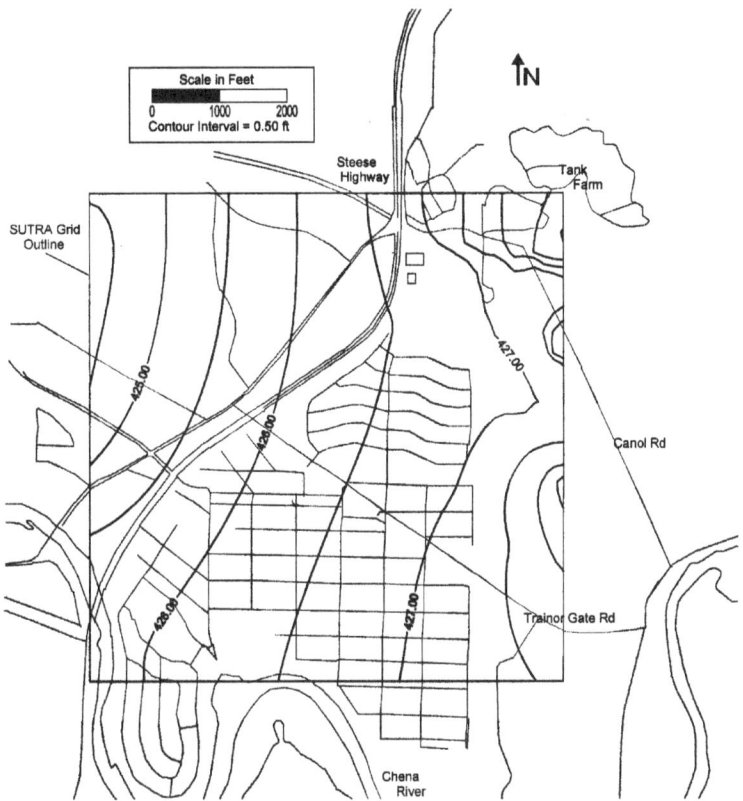

Figure 14.24B Water elevation contours for Tank Farm simulation with permafrost.

Table 14.8 Best-Fit Hydrologic Parameters for
the Tank Farm Simulation

Parameter	Value
K_f	1.00×10^{-6} m/s
K_u	2.70×10^{-3} m/s
Porosity	0.4
Recharge	12.29 cm/year

the hydrology of the system as was seen in Figures 14.24A and 14.24B. If the configuration used in the model does not closely match actual field conditions, the water elevations would be affected. The configuration used to represent permafrost in the model may have some error in it caused by discretizing to the grid elements. The finer detail of the configuration is lost due to the size of the elements. The GPR work done by Lawson et al. (1994) also has some error inherent to it.

Second, there is likely to be some error in the head values used for boundary conditions. The contouring program used to derive head values for the entire grid area does not include any possibility of heterogeneities or abnormalities in the water elevations. It simply interpolates between the measured points, in this case only about 20 points, to get the other values. This presents an error that cannot be removed or checked without additional field measurements.

Third, a steady-state hydrologic model was used. The system may not be at steady state or the measured values may not represent the steady-state condition. The values used represent the 1994 and 1995 conditions. An arithmetic average of these values may not represent the steady-state conditions of the aquifer for the last 50 years or the next 50 years.

Contaminant Transport

Once the best-fit hydrologic parameters were obtained, contaminant transport was added to the simulation. These simulations were also completed with and without permafrost to investigate the effects of permafrost on the contaminant transport pathways.

Contaminant Source

The source for the dissolved contaminant was simulated as a dissolution process from the light NAPLs (LNAPLs). Currier et al. (1994) performed an investigation to delineate the source plume. They discovered a free-phase petroleum plume on the western edge of the Tank Farm (Figure 14.4). They estimated the free-phase plume to be approximately 0.4 ft (0.12 m) thick and 11,200 ft^2 (1040.5 m^2) in area, yielding 4400 gal of free product. A residual saturation of product was indicated by a sheen found on an additional 100,000 ft^2 (9290 m^2) of water. CRREL calculated a residual product volume of approximately 450,000 gal in the soil. The free-phase plume, according to the Lawson et al. (1994) permafrost configuration, was primarily located on shallow permafrost. It was assumed that the only infiltration of the contaminant would come from dissolution into the recharge, which would travel over the shallow permafrost and actually enter the aquifer at the nearest permafrost-free area. It was assumed the recharge would come to equilibrium with the contaminant, allowing the concentration in the recharge to be estimated by the effective solubility of the contaminant. The effective solubility was described in Equation 14.8.

The solubility of pure benzene at 20°C (68°F) is 1780 mg/L (Fetter, 1988). This value was used for the solubility at 4°C (39.5°F) due to limited information on the actual solubility at this temperature. This was considered acceptable as a worst-case scenario, allowing a higher solubility concentration of benzene than may actually occur at 4°C. A benzene mole fraction of 2% by volume in the free-phase plume was obtained from average values listed for percent volume of benzene in gasoline, aviation gasoline, and diesel (Cheremisinoff and Morresi, 1979). The mole fraction initially in the product was calculated along with the effective solubility. An effective solubility of 35.6 mg/L is used as the benzene concentration in the recharge, so the rate of benzene introduction to the aquifer in the simulations is equivalent to the recharge rate. This concentration, 35.6 mg/L, was input into SUTRA as the concentration of benzene in the recharge for the entire duration of the 50-year simulation. No adjustments were made to change the effective solubility, as the mole fraction decreases due to mass transfer to the aquifer or volitization, so this number again represents a worse-case scenario.

Simulation Results

- No Adsorption, No Decay
- Adsorption, No Decay
- Decay, No Adsorption
- Adsorption, Decay

No Adsorption, No Decay. The first transport simulations assume benzene is a conservative chemical. Adsorption on the soil particles and biological decay potential were ignored. These simulations were run to provide a baseline in determining how sensitive the transport was to adsorption or decay. Figure 14.25 illustrates the SUTRA distribution of benzene under these conditions without any permafrost. The

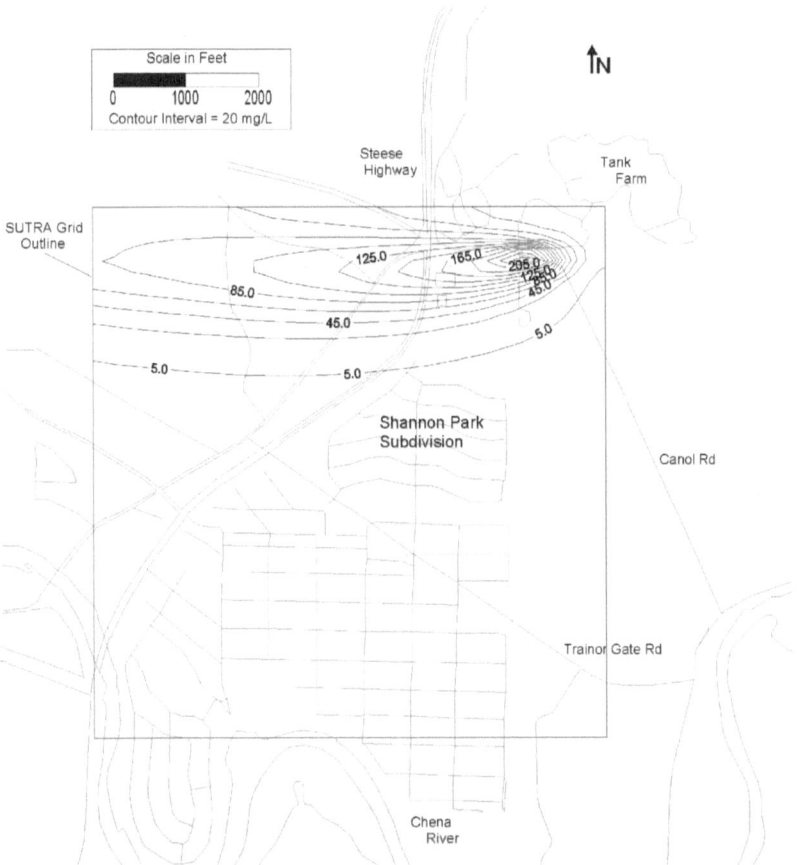

Figure 14.25 Concentration contours for tank farm simulation without permafrost, no retardation factors.

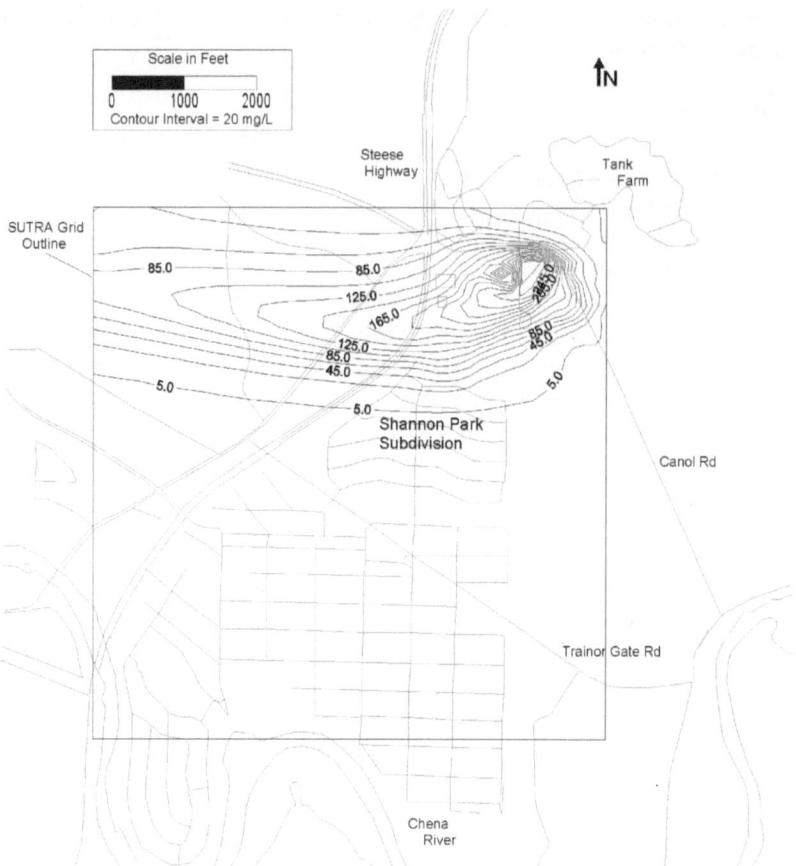

Figure 14.26 Concentration contours for Tank Farm simulation with permafrost, no retardation factors.

contours represent the distribution 50 years after the benzene source was first introduced. Figure 14.26 shows the benzene distribution at the same time step for a simulation where the permafrost is included.

Under these worst-case conditions, the simulations show benzene moving over a mile downgradient. Figure 14.26 shows the effects permafrost has on the system. The two distributions are significantly different. Figure 14.25 illustrates benzene migrating directly west with relatively little movement to the south. Concentrations drop to below the 5 μg/L (MCL for benzene) slightly north of the Shannon Park Subdivision and are approximately 150 μg/L as the plume reaches the Steese Highway. The contours are generally oval shaped, primarily spreading in the direction of the westward advection. The contours in Figure 14.26 are more irregularly shaped and indicate an initial southwest migration of the contaminant. Concentrations do not drop below the MCL until the middle of the subdivision and the concentrations at the Steese Highway are around 185 μg/L. To the north, the concentrations decrease more rapidly than in Figure 14.25. The differences in the two figures fade in the contours to the far west, just as was seen in the water elevational contours. The permafrost seems to have the effect of forcing the water, and thus the contaminant, more to the south. The concentrations indicate that the benzene moves farther through the discontinuous permafrost. This may be a result of the channelization that has formed through the permafrost where the groundwater seems to move faster than the overall hydraulic gradient suggests.

Adsorption, No Decay. The concentrations resulting from the above two simulations in most cases were orders of magnitude higher than those measured in the wells around the Tank Farm facility, and these indicated the contaminant was traveling significantly farther than was indicated by the field measurements, so the next step was to add retardation factors. Adsorption and biodegradation are the most important factors in retarding contaminant movement in the subsurface (Olsen and Davis, 1990), so their effects were investigated with the model. Adsorption was added to the simulation first in the form of a linear Freundlich isotherm. The Freundlich isotherm takes the following form (Fetter, 1988):

$$S = K_d \, C^{(1/N)} \tag{14.17}$$

where

$\quad S \quad$ = concentration of solute adsorbed onto the soil (mass/mass)
$\quad K_d \quad$ = soil water distribution coefficient ($[\text{volume/mass}]^{(1/N)}$)
$\quad C \quad$ = concentration of solute in solution (mass/volume)
$\quad 1/N \quad$ = Freundlich exponent

Olsen and Davis (1990) and Faust and Aly (1987) present evidence that benzene has a Freundlich exponent of 1, which results in a linear isotherm as follows:

$$S = K_d \, C \tag{14.18}$$

The adsorption constant is then the only unknown in the equation.

The soil water distribution coefficient has been studied by several researchers and found to have a strong relationship to the carbon content of the soil as described below (Fetter, 1988; Olsen and Davis, 1990):

$$K_d = 0.63 \, K_{ow} f_{oc} \tag{14.19}$$

where

$\quad K_{ow} \quad$ = octanol water partition coefficient (volume/mass)
$\quad f_{oc} \quad$ = fraction of total organic carbon (TOC) in the soil

The octanol water coefficient for benzene has been measured experimentally and is reported by Olsen and Davis (1990) to be 132 m^3/kg. This value for K_{ow} was combined with the TOC measurements that were taken on the soils in the Tank Farm area to yield a distribution coefficient. The percent TOC ranged from 0.126 to 0.615, which gave a range for K_d of 0.1 to 0.6 m^3/kg. Olsen and Davis (1990) reported values for K_d between 0.02 and 0.09 m³/kg. Faust and Aly (1987) reported values between 0.001 and 1.2 m³/kg.

To cover the wide range of possible values, four separate values were simulated: 0.001, 0.01, 0.06, 0.1 m³/kg; higher values were not realistic in this situation. The trend of retardation was clear for all simulations; less migration is observed with increasing K_d.

A K_d value of 0.1 m³/kg allows less than 200 ft (61 m) of westward migration from the source, and there appeared to be no effect on the migration from the permafrost. With a K_d value of 0.06 m³/kg, there was also no noticeable difference between the simulation with permafrost and that without. The contaminant does migrate approximately 100 ft (30.5 m) farther west in both cases. The values of 0.001 and 0.01 m³/kg allow enough movement of the contaminant to begin to see the effects of the permafrost. With a value of 0.01 m³/kg, the contour patterns for the simulation with permafrost show two distinct pathways and a slight southward migration (Figure 14.27). The contaminant has migrated almost to the Steese High-

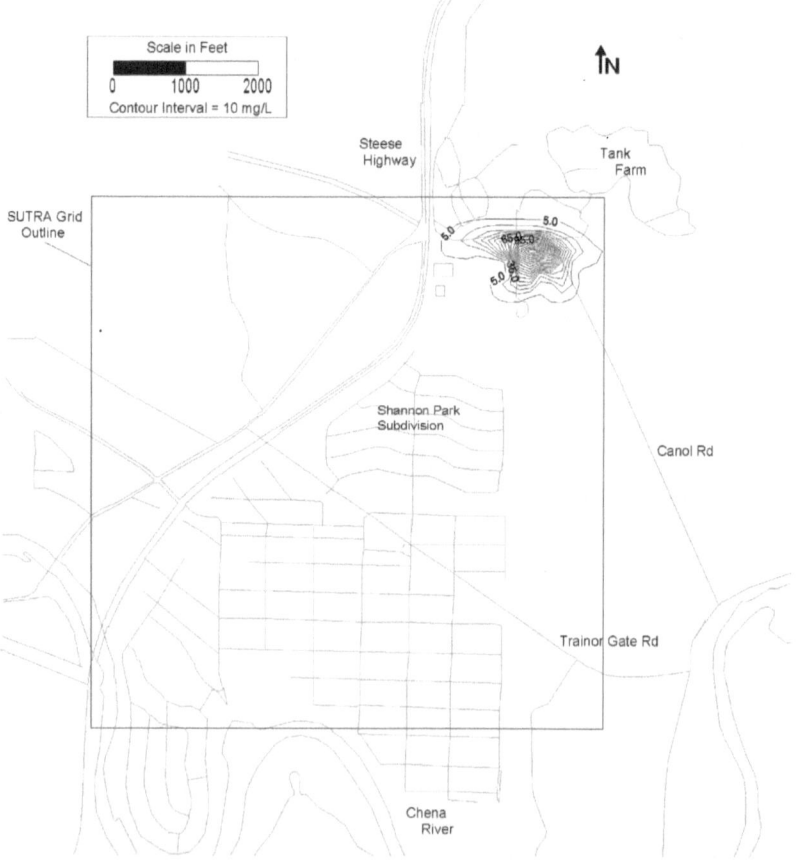

Figure 14.27 Concentration contours for tank farm simulation with permafrost, distribution coefficient = 0.01, no decay.

way (approximately 2000 ft from the source) for both with and without permafrost, but only in specific channels in the simulation with permafrost.

A K_d value of 0.001 m^3/kg allowed the most contaminant migration from the source: 6000 ft through the permafrost, over 6000 ft without the permafrost. The channelization was clearer and the contaminant has more of a southward migration with permafrost included. It is likely that the concentrations to the west are not as high as without permafrost because movement is being directed south. A summary of the SUTRA results at specific well locations is listed in Table 14.9. The measured field concentrations are included for comparison.

Decay, No Adsorption. The second factor included in the simulations was biological decay. First-order decay was assumed for benzene (several researchers have estimated a first-order decay coefficient for benzene [Olsen and Davis, 1990; Ramert and Eberhardt, 1996; Gould and Wallace, 1996; Chapelle et al., 1996]). The equation for first-order decay is given by Drever (1982):

$$C = C_o e^{-kt} \tag{14.20}$$

where

C = concentration (µg/L)
C_o = initial concentration (µg/L)
k = first-order decay coefficient (1/day)
t = time (days)

Literature values for this parameter indicated a range between 0.001 and 0.01 days^{-1} (Olsen and Davis, 1990; Ramert and Eberhardt, 1996; Gould and Wallace, 1996; Chapelle et al., 1996). Values of 0.001, 0.005, 0.01, and 0.1 day^{-1} were simulated with SUTRA to determine the effect this parameter would have on the benzene concentrations.

Figure 14.28 shows the effect of the first-order decay with permafrost included. All figures show benzene concentrations at the 50-year time step. A low rate of 0.001 day^{-1} causes a decrease in benzene concentrations of 100 µg/L from the simulation without any decay. A rate of 0.005 day^{-1} decreases concentrations over

Table 14.9 Benzene Concentrations at Various Well Locations for Simulations with Permafrost, with Adsorption

Well	Field Data July 1994 (µg/L)	SUTRA Concentrations (µg/L) $K_d = 0$ (m^3/kg)	$K_d = 0.001$ (m^3/kg)	$K_d = 0.01$ (m^3/kg)	$K_d = 0.06$ (m^3/kg)	$K_d = 0.1$ (m^3/kg)
5271	92.0	301.5	289.4	91.7	10.4	4.5
5273	144.0	36.8	33.7	1.1	0.0	0.0
5274	0.0	154.5	127.1	6.1	0.0	0.0
5275	0.0	157.8	104.7	0.2	0.0	0.0
6053	0.0	151.5	141.8	30.8	0.8	0.1
6054	1.7	308.0	301.6	184.3	71.8	48.2
6058	0.0	7.2	2.9	0.0	0.0	0.0
6071	2.3	88.7	80.8	6.6	0.0	0.0

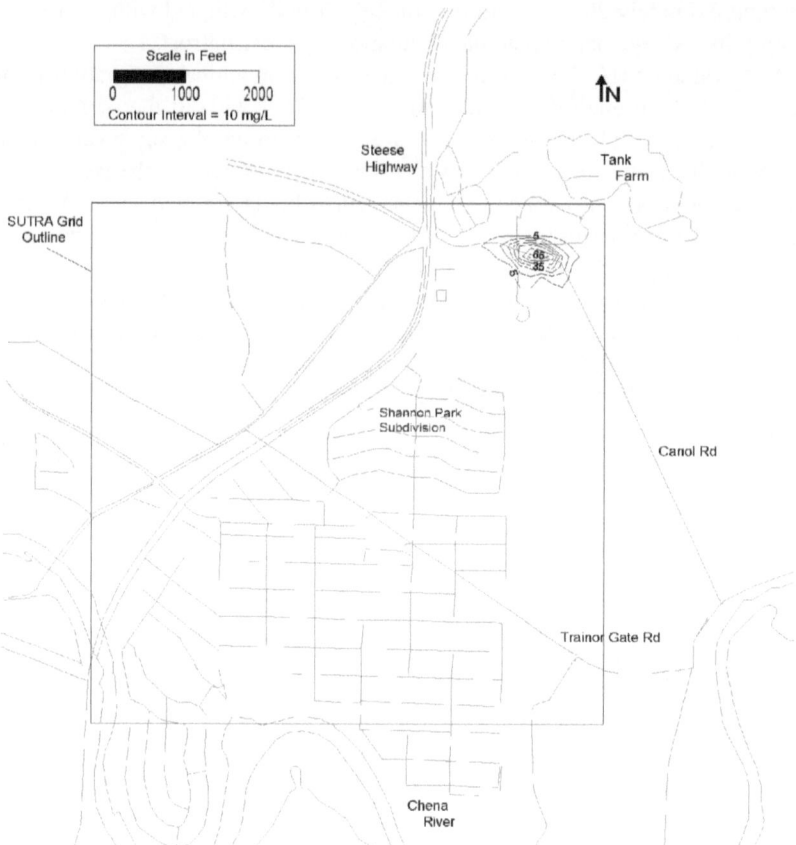

Figure 14.28 Concentration contours for Tank Farm simulation with permafrost, no adsorbtion, first-order decay rate = 0.010 day[1].

160 µg/L and a rate of 0.1 day[-1] decreases the concentrations over 200 µg/L compared with the results from the simulation without any decay.

A decay rate of 0.1 day[-1] shows the benzene being degraded before it is able to migrate beyond the source area. This rate does not appear to be representative of the Tank Farm conditions, as field observations showed the contaminant further downgradient. This simulation does show, however, the effect biodegradation can have on the transport. Each order of magnitude increase in decay coefficient decreased the downgradient concentrations by approximately 100 µg/L. This correlation appeared in simulations including permafrost as well. The channelization effects in the permafrost are seen again in these simulations. Benzene moves downgradient nearly as far as when no permafrost is included, but only in the permafrost-free channels. The preferential southward migration is evident in the first two simulations with decay rates of 0.001 and 0.005 day[-1]. The results for five simulations along with the field data are listed in Table 14.10.

At decay rates of 0.005 day[-1], the set of simulations with and without the permafrost results in approximately the same concentrations at these well locations.

Table 14.10 Benzene Concentrations at Various Well Locations for Simulations with Permafrost, with Decay

Well	Field Data July 1994 (μg/L)	SUTRA Concentrations (μg/L)				
		$K_d = 0.0$ (day⁻¹)	$K_d = 0.001$ (day⁻¹)	$K_d = 0.005$ (day⁻¹)	$K_d = 0.01$ (day⁻¹)	$K_d = 0.1$ (day⁻¹)
5271	92.0	301.5	0.3	0.0	0.0	0.0
5273	144.0	36.8	210.7	111.5	74.2	11.8
5274	0.0	154.5	152.1	43.2	18.8	0.5
5275	0.0	157.8	17.8	0.3	0.0	0.0
6053	0.0	151.5	64.8	13.2	4.2	0.0
6054	1.7	308.0	10.2	0.7	0.1	0.0
6058	0.0	7.2	38.3	3.2	0.6	0.0
6071	2.3	88.7	28.8	3.2	0.6	0.0

At lower decay rates, the simulations without permafrost generally show a lower concentration at the eastern and southern well locations (wells 5271, 5273, 5274, 5275, and 6054) than the simulations with permafrost. The relatively western well locations (wells 6058, 6071, and 6053) had higher concentrations in the simulation without permafrost.

Adsorption, Decay. The final conditions simulated include both adsorption and decay processes to investigate the accumulative effect of the two factors. One set of conditions was simulated, an adsorption coefficient of 0.012 m³/kg with a decay rate of 0.005 day⁻¹. These values were chosen as middle-range values and, from the previous simulations, they seem to best match the field observations. The model results with permafrost included are presented in Figure 14.29. Table 14.11 shows the numerical results for both simulations along with the field data.

The contaminant migrated only 1000 ft downgradient under these conditions before the concentration dropped below the MCL level. Without any retardation factors, benzene traveled over 6000 ft (1829 m) before the concentration dropped below the MCL. Relatively small rate coefficients made a significant impact in the distance of contaminant migration.

CONCLUSIONS

Discontinuous permafrost and other potential heterogeneities in the subsurface system around the Tank Farm area make the hydrology and transport interesting and challenging to simulate with a computer model. Based on the field observations and SUTRA simulations of the area, several conclusions can be drawn. The permafrost clearly affects the water and contaminant migration pathways in the area. The thaw channel running to the north of the Truck Fill Stand was shown by the tracer study data to be a hydrologically significant pathway, flowing east to west. Contours of the water elevations in and around the Tank Farm area indicate localized anomalies in the gradient and flow patterns in the permafrost area. The simulated elevations show a marked difference in the runs with and without permafrost, but also indicate that the influence on the gradients and direction of flow are insignificant 500 ft past the permafrost masses.

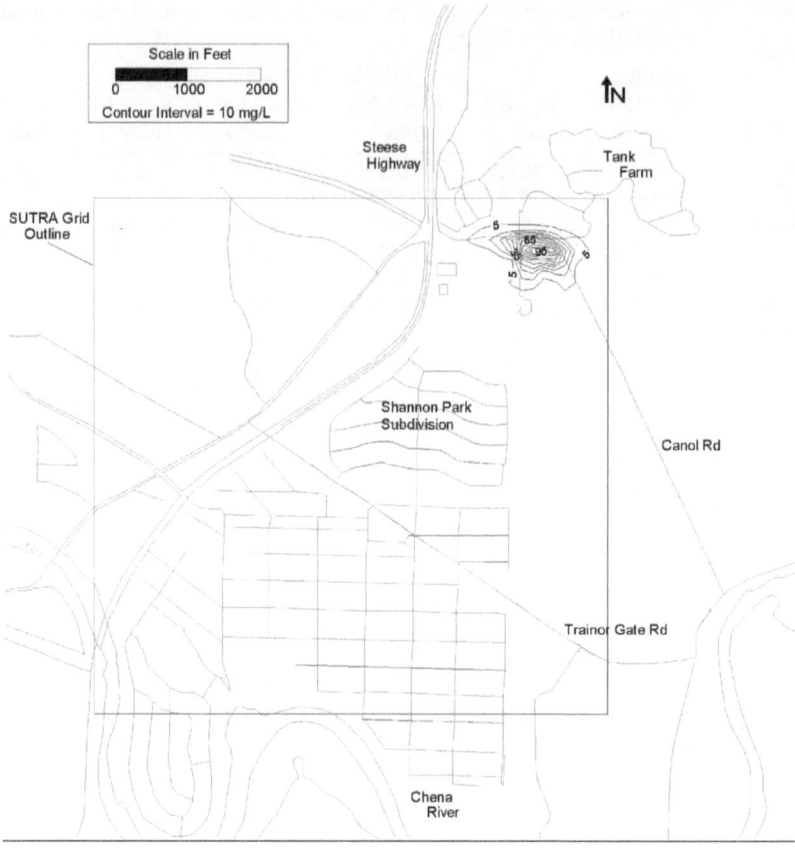

Figure 14.29 Concentration contours for Tank Farm simulation with permafrost, distribution coefficient = 0.012, first-order decay rate = 0.005 day^{-1}.

Table 14.11 Benzene Concentrations at Various Well Locations for Simulations with Adsorption and Decay

	Field Data	SUTRA Concentrations (μg/L)	
Well	July 1994 (μg/L)	With Permafrost	Without Permafrost
5271	92.0	36.5	44.7
5273	144.0	0.258	0.077
5274	0.0	1.48	5.78
5275	0.0	0.032	0.005
6053	0.0	1.0	11.2
6054	1.7	106.8	101.3
6058	0.0	0.000	0.004
6071	2.3	1.71	1.25

The transport simulations show the influence of the permafrost on the migration pathways. The contaminant movement is channeled by the permafrost configuration and directed further south than if there was no permafrost. The downgradient concentration of the benzene is not necessarily diminished, however, by the permafrost masses, but rather more channeled. This would be important in a field investigation with a limited number of wells. If the permafrost configuration is not understood, the contaminant may be "missed" because the transport pathways may be different than that indicated by an areawide hydraulic gradient.

There are notable limitations to these results, most of which could be overcome with further field data and a true three-dimensional computer code. First, the tracer study should be verified by a third injection. Second, both the tracer data and transport simulations should be simulated with a three-dimensional model. Finally, a different set of boundary conditions should be used in the hydrologic simulations to test the accuracy of the boundary conditions used in these simulations.

The results of the tracer simulation are affected by the two-dimensionality of SUTRA. The sensitivity analysis indicated the results are sensitive to the grid thickness, or the third dimension. This is due to the concentrations being averaged over that dimension. A true three-dimensional model would allow the concentration to vary in the third dimension. A three-dimensional code might also allow a better calibration of the areawide transport simulations with the existing field data. Measured concentrations at various depths below the water table could be compared directly with results from a three-dimensional simulation. The results could then be statistically compared with the measured field data.

REFERENCES

APHA, AWWA, and WPCF. Standard Methods for the Examination of Water and Wastewater, 18th ed. American Public Health Association, Washington, D.C., 1992.

Bear, J. *Hydraulics of Groundwater*. McGraw-Hill, New York, 1979, 569 pp.

Bolton, B. Personal communication, USGS Hydrologist, Fairbanks, AK, May 1996.

Chapelle, F.H., P.M. Bradley, D.R. Lovley, and D.A. Vroblesky. Measuring Rates of Biodegradation in a Contaminated Aquifer Using Field and Laboratory Methods, *Groundwater*, 34(4), 691–698, 1996.

Cheremisinoff, P.N. and A.C. Morresi. *Benzene: Basic and Hazardous Properties*. Marcel Dekker, New York, 1979, chap. 1.

Claar, D. and B. Bolton. Personal communication, USGS Hydrologists. Fairbanks, AK, March 1996.

Claar, D. Personal communication, USGS Hydrologist. Fairbanks, AK, September 1995.

Currier, P.M. and D.C. Leggett. Quality Assurance Program Plan, U.S. Army Cold Regions Research and Engineering Laboratory, Hanover, NH, 1994.

Currier, P.M., B.D. Taras, and P.R. Peapples. Supplemental Source Delineation and Groundwater Flow Investigation in the Vicinity of the Tank Farm Area Building 1173 Using Microwells, Preliminary Report. U.S. Army Cold Regions Research and Engineering Laboratory, Geochemical Sciences Branch, Hanover, NH, 1994, 15 p.

Dingman, S.L. Hydrologic Studies of the Glenn Creek Drainage Basin near Fairbanks, Alaska. Research Report 297, U.S. Army Cold Regions Research and Engineering Laboratory, Hanover, NH, 1971, 112 p.

Drever, J.I. *The Geochemistry of Natural Waters*. Prentice-Hall, Englewood Cliffs, NJ, 1982, 388 p.

Ecology and Environment, Inc. Final Management Plan, Operable Unit 3, Fort Wainwright, Alaska. Report prepared for U.S.A.C.O.E. and 6th ID Light, Anchorage, AK, 1993.

Ecology and Environment, Inc. Remedial Investigation Report, Operable Unit 3, Vol. 1 and 3, Report prepared for U.S.A.C.O.E. and 6th ID Light, Anchorage, AK, 1994.

Elprince, A. and P. Day. Fitting Solute Breakthrough Equations Using Two Adjustable Parameters, *Soil Sci. Soc. Am. J.*, 41, 39–41, 1977.

Farris, A.M. Numerical Modeling of Contaminant Transport in Discontinuous Permafrost — Fort Wainwright, Alaska. M.S. thesis. University of Alaska, Fairbanks, 1996, 109 pp.

Faust, S.D. and O.M. Aly. *Adsorption Processes for Water Treatment*. Butterworths Publishers, Boston, 1987, 509 p.

Federal Register. Vol. 49, no. 2. U.S. Government Printing Office, Washington, D.C., 1984.

Fetter, C.W. *Applied Hydrogeology*. Macmillan, New York, 1988, 591 p.

Freeze, R.A. and J.A. Cherry. *Groundwater*. Prentice-Hall, Englewood Cliffs, NJ, 1979, 604 p.

Freyberg, D. A Natural Gradient Experiment on Solute Transport in a Sand Aquifer. II. Spatial Moments and the Advection and Dispersion of Nonreactive Tracers, *Water Resources Research*, 22, 2031–2046, 1986.

Fried, J. *Groundwater Pollution*. Elsevier, New York, 1975.

Garabedian, S.P., D.R. LeBlanc, L.W. Gelhar, and M.A. Celia. Large-Scale Natural Gradient Tracer Test in Sand and Gravel, Cape Cod, Mass. II. Analysis of Spatial Moments in a Nonreactive Tracer, *Water Resour. Res.*, 27, 911–924, 1991.

Gelhar, L.W., C. Welty, and K.R. Rehfeldt. A Critical Review of Data on Field-Scale Dispersion in Aquifers, *Water Resour. Res.*, 28(7), 1955–1974, 1992.

Gieck, R.E., Jr. and D.L. Kane. Hydrology of Two Subarctic Watersheds, in *Proceedings of the Cold Regions Hydrology Symposium*. D.L. Kane, Ed. American Water Resources Association, Fairbanks, AK, 1986, 283–291.

Gould, T.F. and M. Wallace. A Pilot-Scale Study of *In Situ* Hydrocarbon Remediation of Contamination in Soil and Groundwater at Fort Wainwright, Alaska, in *Cold Regions Engineering, Proceedings of the Eight International Conference on Cold Regions Engineering*, ASCE, New York, 1996, 106–115.

Kane, D.L., R.E. Gieck, and L.D. Hinzman. Evapotranspiration from a Small Alaskan Arctic Watershed, *Nordic Hydrol.*, 21(4/5), 253–272, 1990.

Kane, D.L. and J. Stein. Field Evidence of Groundwater Recharge in Interior Alaska, in *Permafrost Fourth International Conference Proceedings*. National Academy Press, Fairbanks, AK, 1983a, 572–577.

Kane, D.L. and J. Stein, Water Movement into Seasonally Frozen Soils, *Water Resour. Res.*, 19(6), 1547–1557, 1983b.

Keckler, D. *SURFER for Windows User Manual, Version 6*. Golden Software, Inc., Golden, CO, 1995.

Lawson, D.E., J.C. Strasser, and J.M. Davi. Geological and Geophysical Investigations of the Hydrogeology of the Operable Unit 3, Interim Draft Report. U.S. Army Cold Regions Research and Engineering Laboratory. Prepared for U.S. Army Engineers District, Alaska and U.S. Department of Army 6th Infantry Division, Fort Richardson, 1994, 28 p.

Light, G. An Assessment of Benzene Concentration Variations in Groundwater Samples Extracted from Monitoring Wells. M.S. thesis. University of Alaska Fairbanks, 1996, 136 pp.

Lilly, M.R. and S.R. Ray. Late-Winter Chemical and Physical Characteristics of the Fairbanks Alluvial Aquifer, *Water Quality in the Great Land — Alaska's Challenge, Proceedings, American Water Research Association*, Water Research Center, Institute of Northern Engineering, Report IWR-109, University of Alaska, Fairbanks, 1987, 127–140.

McCort, D. Personal communication, Fairbanks Fuel Terminal Manager, Fairbanks, AK, July 1994.

Mueller, S.W. Permafrost or Permanently Frozen Ground and Related Engineering Problems. U.S. Army Office Chief of Engineers, Military Intelligence Division Strategic Engineering Study 62, 1943, 231 p.

National Climatic Data Center. 1995. Meteorological Data for 1995, Fairbanks, Alaska. National Climatic Data Center, Asheville, NC, 1995, 6 p.

Natural Resources Conservation Service. Alaska Annual Data Summary. U.S. Department of Agriculture, Anchorage, AK, 1994 and 1995.

Olsen, R.L. and A. Davis. Predicting the Fate and Control of Organic Compounds in Groundwater, *Hazardous Materials Control*, 3(3), 39–64, 1990.

Plumb, E.W. and M.R. Lilly. Snow-Depth and Water-Equivalent Data for the Fairbanks Area, Alaska, Spring 1995. Prepared in Cooperation with the Fairbanks International Airport; University of Alaska, Fairbanks; Alaska Department of Natural Resources, Division of Mining and Water Management; U.S. Army, Alaska; U.S. Army Corps of Engineers, Alaska District, OF 96-0414, ALASKA, 16 p.

Priestley, C.H.B. and R.J. Taylor. On the Assessment of Surface Heat Flux and Evaporation Using Large-Scale Parameters, *Monthly Weather Rev.*, 100, 81–92, 1972.

Ramert, P.C. and W.L. Eberhardt. Petroleum Hydrocarbon Removal via Volatilization and Biodegradation at McGrath, Alaska, in *Cold Regions Engineering, Proceedings of the Eight International Conference on Cold Regions Engineering*. ASCE, New York, 1996, 106–115.

Rouse, W.R. and R.B. Stewart. A Simple Model for Determining Evaporation from High Latitude Upland Sites. *J. Appl. Meteorol.*, 11, 1063–1070, 1972.

Rovansek, R.J., D.L. Kane, and L.D. Hinzman. Improving Estimates of Snowpack Water Equivalent Using Double Sampling, Eastern Snow Conference, Quebec, Canada, 1993.

Stewart, R.B. and W.R. Rouse. Simple Models for Calculating Evaporation from Dry and Wet Tundra Surfaces. *Arctic Alpine Res.*, 8(3), 263–274, 1976.

U.S. Army Corps of Engineers, Groundwater Monitoring Network, Fort Wainwright, Alaska. U.S. Army Installation Restoration Program, Alaska District, 1993.

Van Genuchten, M. Analytical Solutions for Chemical Transport with Simultaneous Adsorption, Zero-Order Production and First-Order Decay, *J. Hydrol.*, 49, 213–233, 1981.

Voss, C.I. SUTRA-Saturated Unsaturated Transport Manual. U.S. Geological Survey Water Resources Investigations Report 84–4369, Denver, CO, 1984, 409 p.

Index

239